Exercises in Plant Physiology

Second Edition

Exercises in Plant Physiology

Second Edition

Francis H. Witham
The Pennsylvania State University

David F. Blaydes
West Virginia University

Robert M. Devlin
University of Massachusetts

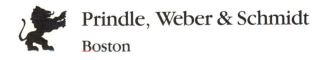
Prindle, Weber & Schmidt
Boston

PWS PUBLISHERS

Prindle, Weber & Schmidt • ☙ • Duxbury Press • ♠ • PWS Engineering • △ • Breton Publishers • ⚙
Statler Office Building • 20 Park Plaza • Boston, Massachusetts 02116

PWS Publishers is a division of Wadsworth, Inc.

Library of Congress Cataloging-in-Publication Data

Witham, Francis H., 1936–
  Exercises in plant physiology.

  Previous ed. published as: Experiments in plant physiology.
  Bibliography: p.
  Includes index.
  1. Plant physiology—Laboratory manuals. I. Blaydes, David F. II. Devlin, Robert M.
III. Witham, Francis H., 1936– . Experiments in plant physiology. IV. Title.
QK714.5.W58 1986   581.1'07'8   85-12052
ISBN 0-87150-944-X

ISBN 0-87150-944-X

Printed in the United States of America
1 2 3 4 5 6 7 8 9—90 89 88 87 86

Sponsoring editor: Jean-Francois Vilain
Editorial assistants: Tyrel Holston and Traci Sobocinski
Production coordinator: Susan Graham
Production: Technical Texts, Inc.
Interior and cover design: Susan Graham
New interior illustrations: Julia Gecha
Composition: A&B Typesetters, Inc.
Cover printing: New England Book Components
Text printing and binding: Halliday Lithograph

Preface

The exercises in the second edition of this manual are intended for use in a one- or two-term course in introductory plant physiology. In order to provide maximum flexibility for individual selection, course design, and available facilities, we offer a large number of exercises that not only vary in content and complexity but also present a broad spectrum of basic physiological concepts and techniques. We chose exercises that work and teach—from classic experiments to modern ones—with the expectation that students who perform them will gain a solid grounding in plant physiology. Ten exercises are new to this edition, and the majority of the others have been revised and fine-tuned.

Although we grouped the exercises according to similarities in subject material, they need not be performed in the sequence they appear in this manual. The integration of selected exercises within the framework of various courses will vary considerably and is left to the discretion of the individual instructor. In this respect, the aids for scheduling exercises and the instructions for preparing specific reagents, which appear in Appendix I and II, respectively, will be of considerable value.

Each exercise contains the following elements: an introduction giving the necessary background information, a concise statement of purpose, a list of materials to be used, a detailed discussion of procedure, a section on results and conclusions, and a list of acknowledgments and references. This organization, aside from solving the mechanics for the presentation of each exercise, illustrates the basic logic and format of scientific communication.

In the materials section, the amount and nature of the materials to be used are indicated for the performance of an entire experiment by a group of students whose number is recommended in the aids for scheduling appendix. From our experience, it appears decidedly advantageous to prepare materials and reagents before the class meeting. However, certain preparatory techniques may be useful to students with minimal backgrounds in chemistry.

The procedure section of each exercise consists of different activities that may or may not be completed. Their completion depends upon the exercise itself and the extent of coverage desired.

With respect to the results obtained, sufficient space or tables are provided within the text for recording data. We suggest that students be encouraged to present their organized and/or illustrated results and conclusions in a brief report, which is an effective means for reinforcing basic concepts and techniques.

Included at the back of the manual are appendixes that provide useful review. The aids for scheduling exercises and the preparation of materials and reagents have already been mentioned. Also included is information on atomic weights, common ions and organic functional groups, chemical reagents and solutions of acids and bases, the Greek alphabet, physical constants and conversion factors, numerical equivalents, and stopper sizes. Finally, a supplementary readings list provides a source for additional background information.

For the most part, the exercises in this manual have been modified for class use from the published creative work of others. We acknowledge this work, but at the same time we assume full responsibility for errors in presenting this material. In addition, the mention of various trade names and suppliers of chemicals and equipment is for convenience only and is not to be construed as an endorsement.

We are indebted to Drs. Carlos O. Miller, Clifford J. Pollard, Charles W. Hagen, Felix Lukezic, and Robert H. Hamilton for their interest at the early stages of this venture and to Drs. Bernard S. Meyer, Donald B. Anderson, and Carroll A. Swanson for the use of those experiments that appeared in their manual *Laboratory Plant Physiology*.

We extend special thanks to Dr. John T. Barber, Tulane University; Dr. Edward A. Funkhouser, Texas A & M University; Dr. Chris K. Kjeldson, Sonoma State University; Dr. Robert B. McNairn, California State University at Chico; Dr. Stephen H. Scheck, Loyola Marymount University; and Dr. Michael S. Strauss, Northeastern University.

We also wish to thank Dr. Sophia B. Blaydes and Mrs. Rose Mary Witham for their assistance and support.

To the student

Before the performance of any experiment, learn the general and specific rules of laboratory safety as outlined by the instructor. Treat all chemicals and reagents as poisonous and/or toxic and caustic unless contraindicated. Also, become familiar with the laboratory as to the location of exits, showers, drinking fountains, and so on. Above all, report all accidents immediately.

Some of the experiments require the use of various instruments that should only be operated according to specific directions. If at any time you are not sure of their proper operation, do not hesitate to seek the aid of your instructor since wasted time and possible damage to the equipment can be easily avoided with suitable precautions.

The following exercises illustrate a variety of techniques and basic concepts that will add appreciably to your understanding of modern plant physiology. The success of your laboratory experience, however, will depend in large part upon your preparation before the regularly scheduled class meeting, the maintenance of accurate records of the experimental data, and the presentation of results and conclusions in a concisely written report.

The purpose and procedural details of each experiment should be read and thoroughly understood before class. With suitable preparation you will be able to devote maximum effort to the required laboratory operations and gain considerable comprehension of the techniques and physiological concepts illustrated. Conversely, waiting to read an assigned exercise during valuable laboratory time will lead to considerable confusion and at best a "cookbook" experience in plant physiology.

It is equally important to record your observations and data in the spaces or tables provided within this manual or in a suitably bound laboratory notebook. Similarly, every attempt should be made to avoid recording the data on a loose piece of paper, which can be easily lost. In addition, even though the exercises in this manual have been completed successfully in the past, there is no guarantee that they will be performed in the same manner by everyone. Therefore, "textbook" results may not be obtained in all instances. Nevertheless, the results should be reported as accurately as possible, for it is evident that the progress of science is based on "calculated trial and error" with intellectual honesty being paramount.

Since the purpose, materials, and procedural details are given for each exercise, it is not necessary to reproduce these sections in a written report. You are encouraged, however, to present your organized results or observations in a suitably designed table or graph. For some experiments, graphs or tables are provided as examples that can be used for other experiments as well. Also, in presenting your conclusions, pay particular attention to the guidelines or questions indicated. Remember that if the results are not as you expected, make note of that fact and offer any explanations that seem applicable. The references at the end of each exercise and the supplementary readings at the end of the manual will provide additional background information and may be used in concluding remarks. Further, the references are provided so that you will become familiar with the supporting literature and obtain sufficient insight into the creative work of those scientists responsible for the contents of this manual.

Contents

EXERCISE 1

Acids and bases

Introduction

An acid is any molecule or ion that donates a proton (H^+) to other molecules or ions. A base, on the other hand, is a molecule or ion that will accept a proton. The ease with which an acid yields a proton through ionization is a measure of its strength. A measure of the strength of a base, however, is the ease with which it accepts protons: A strong base accepts protons readily, while a weak base shows only a weak affinity for protons.

The acidity or basicity of a solution is determined by its hydrogen ion concentration, which is expressed as the solution's negative logarithm or pH value:

$$pH = -\log (H^+)$$

The hydrogen ion concentration of a liter of pure water is 0.0000001N or 10^{-7}. Since the pH is equal to the negative logarithm of the hydrogen ion concentration, then for water:

$$pH = -\log 10^{-7} = \log \frac{1}{10^{-7}} = 7$$

Values below a pH of 7 indicate acid solutions; those above 7 indicate basic solutions. Pure water, with a pH of 7, is considered neutral. Thus, a solution with a pH of 9 has a hydrogen ion concentration one hundred times less than a solution with a pH of 7 and is considered basic.

The pH of cell solutions is of vital importance to the regulation of biochemical and physiological processes in plants. The metabolic activities of plant cells produce a variety of acids and bases, the levels of which must be regulated for plant growth, development, and even survival. Much of the pH regulation of cell solutions depends upon the buffering action of organic acids, amino acids, proteins, and basic substances such as phosphate (HPO_4^{-2}) and bicarbonate (HCO_3^{-2}) ions.

Almost complete ionization takes place when either a strong acid or a strong base is dissolved in water. In contrast, only slight ionization takes place when either weak acids or bases are dissolved in water. For example, the pH of 1N HCl is 0.1, indicating almost complete ionization. The pH of 1N acetic acid, however, is

1

2.4, showing a higher value (less acid) because there is less dissociation of acetic acid and fewer protons contributed to the solution.

Even though there is a vast array of naturally occurring acids and bases, the pH of plant cells or portions of plant cells may be held within a relatively narrow range due to buffering action. But, since the pH of plant cells may influence absorption, ion exchange, transpiration of nutrients, and the rates of enzymatically catalyzed reactions of metabolism, it is essential that plant scientists understand the activity of hydrogen in plant systems.

Purpose

To familiarize the student with pH determination, titration techniques, and buffer action.

Materials

22 test tubes	2 oranges
2 volumetric pipets (10 ml)	1 qt. grapefruit juice
2 volumetric pipets (5 ml)	1 red cabbage
2 Erlenmeyer flasks (125 ml)	acetic acid (0.1N)
2 beakers (50 ml)	hydrochloric acid (0.1N)
1 burette (50 ml)	NaOH (0.1N)
1 funnel to fit burette	phenolphthalein
filter paper (Whatman No. 1)	thymol blue indicator
pH meter	NaCl (1M)
universal indicator	sodium acetate (1N)
pH paper	HCl (1N)
2 lemons	

Procedure

A. Sample preparation

Before preparing the samples for testing, filter the orange and grapefruit juices to ensure that they are free of solids and relatively clear.

Pipet 5 ml aliquots of each of the following materials into separate test tubes: lemon juice, grapefruit juice, orange juice, 0.1N acetic acid, 0.1N HCl, 0.1N NaOH, tap water, and distilled water. Prepare eight additional tubes in the same manner so as to have a duplicate for each sample.

B. Color indicator method

Add two drops of universal indicator to each of the first eight tubes prepared and mix thoroughly. Using a color chart provided by the instructor, compare the colors of the samples with those on the chart. Determine the pH values by finding the closest match between the color of each sample and those on the chart.

C.   **Electrometric method**

The instructor will demonstrate careful use of the pH meter. Standardize the meter by using an appropriate buffer close to the range of the pH values to be measured. After each standardization, wash the electrode free of buffer with distilled water in a "squeeze" bottle and wipe dry with a paper towel.

Measure the pH of each of the remaining eight samples, cleaning the electrodes with distilled water between each measurement. After recording the readings, clean the electrodes and return them to the buffer. Do not discard the samples. If time permits, determine the pH of each of the samples by using pH paper or by a method suggested by the instructor. Construct a table comparing the pH values obtained with the standard electrometric method with the color indicator method.

D.   **Titration**

Pour 50 ml of 0.1N NaOH through a funnel into a clean, dry burette (the stopcock should be greased and closed). Pipet 10 ml of 0.1N HCl to a 125 ml Erlenmeyer flask. Then add five drops of phenolphthalein indicator to the flask and mix. Record the level of the meniscus in the burette and begin adding 0.1N NaOH a drop at a time to the acid in the flask. After each drop of base, mix the contents by rotating the flask. Continue adding the NaOH to the acid until a faint pink color remains after thorough mixing. The acid is now neutralized. Record the amount of base added by subtracting the initial reading of the meniscus in the burette from the final reading.

A small square of paper with an accurately drawn straight line may help you read the level of the meniscus of the liquid. A sheet of white paper held beneath the flask can help you identify the solution's pink color, which indicates the end point of the titration. Repeat the titration using 10 ml of 0.1N acetic acid.

Pipet 5 ml of 0.1N acetic acid into one test tube and 5 ml of 0.1N hydrochloric acid into another. To each of the test tubes, add five drops of the indicator thymol blue. Record the color. What does it indicate?

Titration of plant juice

Following the same general procedure, select one of the plant juices and titrate it from pH 3.0 to 11.0 using 0.1N NaOH and 0.1N NHCl.

E.   **Buffers**

Using the electrometric method of measuring pH described above, determine the pH of a 1M NaCl solution. Pipet 10 ml of the 1M NaCl into a 50 ml beaker and then add 1 ml of 1N HCl. Stir and determine the pH. Why did the pH change?

Again using the electrometric method for measuring pH, determine the pH of a 1N sodium acetate solution. Then add 1 ml of 1N HCl to 10 ml of the sodium acetate; stir, and determine the pH. Was the change in pH after the addition of the HCl greater with the sodium acetate or with the NaCl solution? Why? What is meant by buffer action?

F.   Natural indicators

Mince about 25 g of red cabbage tissue and add 100 ml of distilled water. Slowly heat the solution to 80°C and keep the temperature constant, stirring until it is red. Filter, then add 5 ml of the solution to each of two test tubes. To one of the tubes, add several drops of 0.1N NaOH. Observe the change in color. To the other test tube, add several drops of 0.1N HCl. Again observe the color change. What are the anthocyanins? What is their location within the cell? What are their chemical properties?

Results/
Conclusions

In concluding statements, answer the following:

1.   What is the meaning of total acidity? What does the pH mean? Why is it incorrect to speak of an "average" pH value?

2.   What are the pH values of the following HCl solutions (assume 100% dissociation): 1N, 0.1N, 0.01N, 10^{-4}N, 10^{-5}N? Why is the pH of a 10^{-8}N HCl not pH 8?

3.   What is the concentration of H^+ ions in water? What is the neutral pH? Why is the pH of distilled water sometimes below 7?

4.   What does titration measure? What do the indicators measure?

References

Devlin, R. M. and F. H. Witham. 1983. *Plant Physiology*, 4th edition. Boston: Willard Grant Press.

Giese, A. C. 1979. *Cell Physiology*, 5th edition. Philadelphia: W. B. Saunders Co.

Raven, P. H., R. F. Evert, and H. Curtis. 1981. *Biology of Plants*, 3rd edition. New York: Worth Publishers, Inc.

White, E. H. 1970. *Chemical Background for the Biological Sciences*, 2nd edition. Englewood Cliffs, N. J.: Prentice-Hall, Inc.

EXERCISE 2

Diffusion

Diffusion is the net movement of a substance from an area of its own high chemical potential into another area of lower chemical potential. It is dependent on a chemical potential gradient that becomes less steep as the process proceeds. When the gradient no longer exists or when equilibrium is attained, diffusion will cease. Any factor that influences the chemical potential gradient will similarly influence the diffusion process. Such factors include temperature, the density of the diffusing molecules, their solubility in the diffusion medium, and the chemical potential gradient. Factors that control the rate of diffusion of gases will also largely control the rate of diffusion of liquids and solids as well.

Two additional factors that influence the diffusion of solutes in solvents, liquids in liquids, and gases in liquids are the size of the diffusing molecules and their solubility in the diffusion medium. The rates of diffusion of gases under constant conditions (see Procedure, Part A) are inversely proportional to the square roots of their relative densities, as shown in the following relationship:

$$\frac{r_1}{r_2} = \frac{\sqrt{d_2}}{\sqrt{d_1}}$$

where r_1 and r_2 are the diffusion rates of the gases that have the densities d_1 and d_2, respectively.

With the application of NH_4OH and HCl to cotton on opposite ends of a glass tube, NH_3 and HCl will diffuse toward each other at rates dependent on the mass of their molecules. The location where they meet will be indicated by a resulting white cloud of ammonium chloride. A measure of the distance traveled by each gas will give a relative measure of their respective rates of diffusion.

The remaining parts of this exercise are designed to illustrate the comparative rates of diffusion of a gas through a gas or liquid medium. In that portion of the exercise on diffusion of solute, remember that when I_2KI solution reacts with starch, you will see its color change to dark blue. The time for the color change (due to the iodine–starch reaction) after the initial setup will indicate the relative rate of solute movement.

Purpose

To study the process of diffusion and some of the factors that influence this phenomenon.

Materials

HCl (concentrated) in dropping bottle

NH$_4$OH (concentrated) in dropping bottle

NaOH (0.1N) in dropping bottle

50 ml methyl red solution (see Appendix II)

10 ml chloroform

10 ml eosin solution (1%, w/v)

5 ml xylol

5 ml ethyl ether

200 ml starch solution (2%, w/v)

200 ml I$_2$KI solution (see Appendix II)

2 portions of water (each 500 ml at different temperatures: 5 to 10°C, 25°C)

glass tube containing a gel of solidified bacto-agar and methyl red adjusted to alkaline color with NaOH (see Procedure, Part B and Figure 2–1 for preparation)

glass tube containing a strip of filter paper moistened with methyl red indicator adjusted to alkaline color (see Procedure, Part B and Figure 2–1 for preparation)

2 small bottles or jars corked with a one-hole stopper and containing concentrated HCl (see Figure 2–1)

glass tube (½ in. diameter and 16 in. long)

4 ring stands and clamps

cotton

2 test tubes

6 strips dialysis tubing (7 in. long, soaked in water)

12 screw clamps

3 plastic trays (7 in. by 4 in.)

Procedure

A. Simple diffusion of gases

Support a glass tube (approximately ½ in. in diameter and 16 in. long) parallel to the bench top with two ring stands and clamps. Plug each end of the tube with a wad of cotton.

To begin the experiment, remove the plugs and place three to five drops of concentrated HCl in one cotton plug and three to five drops of concentrated NH$_4$OH on the other plug. Then reinsert the plugs, introducing the ends to which the acid or base was applied into the tube. Be sure to place the plugs into opposite ends of the tube at the same time.

Note the time it takes for the appearance of a narrow white ring in the tube. Immediately on observing the ring, mark its position and measure the distance between the ring and each end of the tube. Repeat the experiment several times. Be sure to wash and thoroughly dry the tube before each run.

Measurements from each end of the tube:

Indicate the specific gases that have diffused through the tube and write the equation illustrating the reaction involved in the formation of the ring.

Calculate the relative rates of diffusion for the gases in question and determine how much faster the lighter gas diffuses according to the following equation:

$$\text{relative rates of diffusion} = \frac{\text{distance traveled by faster diffusing gas}}{\text{distance traveled by slower diffusing gas}}$$

Now compare the relative densities of the gases in question as an inverse relationship to the relative rates of diffusion according to the following equation:

$$\frac{\text{rate of faster diffusing gases}}{\text{rate of slower diffusing gases}} = \frac{\sqrt{\text{heavier gas (mol wt)}}}{\sqrt{\text{lighter gas (mol wt)}}}$$

According to the preceding equation (based on Graham's law of diffusion), the value obtained will indicate how much faster the lighter gas diffuses than the other. Are your experimental results in accordance with Graham's law?

B.   Comparative rates of diffusion through gas and liquid

Add 2 g of bacto-agar to 70 ml of water. Melt the agar and add 10 ml of methyl red indicator (see Appendix II) and a sufficient number of drops of 0.1N NaOH to adjust the indicator to its yellow or alkaline color. Then adjust the volume of the melted agar mixture to 100 ml with hot water. Fill a glass tube (approximately 1 cm internal diameter, 10 cm long, and stoppered at the bottom) with the melted agar mixture.

When the agar has solidified, support the tube with a ring stand and clamp with the open end down over a small bottle containing concentrated HCl (Figure 2–1). Measure the distance to the diffusion front at suitable time intervals and record the rate of diffusion in millimeters per hour. Since the agar will not appreciably retard solute diffusion, it may be assumed that the diffusion medium is close to that of water.

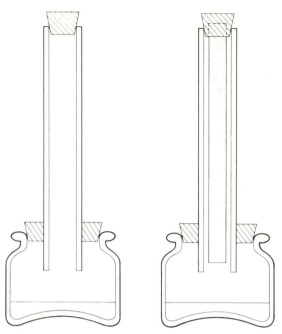

A. Tube filled with Agar Gel
   Colored with Methyl
   Red Indicator

B. Tube Enclosing Filter Paper
   Strip Moistened with Methyl
   Red Indicator

FIGURE 2–1   Comparative rates of diffusion through gas and liquid. (Used with permission of B. S. Meyer, D. B. Anderson, and C. A. Swanson. 1955. *Laboratory Plant Physiology,* 3rd edition. D. Van Nostrand Co., Inc., Princeton, N.J., p. 18.)

As indicated in Figure 2–1, cut a strip of filter paper (slightly less than 1 cm in diameter and 10 cm long) and soak it in methyl red indicator (stock solution adjusted to a yellow color with a little 0.1N NaOH). Suspend the wet strip in a second tube (1 cm diameter and 10 cm long). Support the tube over a small bottle of HCl.

Record the rate of diffusion of HCl through a gaseous medium in millimeters per hour.

Observations:

C.   Diffusion of liquids

Place 5 ml of chloroform in a test tube and on top of the chloroform slowly add sufficient water (colored with eosin) until a colored layer about 5 mm deep is obtained. Then carefully add 5 ml of ether and stopper the tube tightly. Do not shake the tube.

Using a second test tube perform the same procedure, but substitute xylol for the ether as the upper layer of liquid.

Carefully mark the interfaces between the layers of each tube and set the tubes aside, taking precautions to avoid jarring. Examine the tubes several times during the week and record the changes in the position of the interfaces (changes in position of the eosin-water layer).

Observations:

D.   Diffusion of solute

Select a strip of dialysis tubing (about 7 in. long) that has been presoaked in water. Twist one end, bend it back on itself, and then secure it tightly with a screw clamp. Open the other end by rubbing it between your finger and fill it to within one inch of the top with starch solution. Now secure the open end in the same manner. Be sure the clamps are tight and that there are no leaks. Wash the tubing to remove any starch on the outer surface. Place the tube in a plastic tray (about 7 in. by 4 in.) along the long side. Add water at 5 to 10°C (use refrigerated water or water cooled with ice) until it just barely covers the tubing.

Select another piece of dialysis tubing and fill it with potassium iodide iodine solution (I_2KI). Clamp both ends securely. Place this tube in the tray of water, but along the opposite side of the tray. Do not allow the tubes to come into contact with each other. If there are any leaks they will be observed immediately and should be corrected.

Set up another tray as before, but the temperature of the water in this tray should be about 25°C.

In all cases, record the time after both tubes have been placed in the water; over the next two hours frequently observe any volume or color changes in the tubes and in the water in the trays.

Observations:

Results/
Conclusions

The results of each part may be illustrated by tables based on your observations. According to the circumstances, indicate which molecules diffuse, the direction of diffusion, and the nature of the medium through which diffusion occurred.

In concluding statements, answer the following:

1.   What is Graham's law of diffusion, and how does it apply to the diffusion of gases?

2.   What is Fick's law of diffusion, and how does it apply to those factors that govern the diffusion of solutes?

3.   What does the term *diffusion coefficient* mean?

4.   What factors affect the diffusion rate of gases and solutes, and has the effect of some of those factors been demonstrated in the various parts of this exercise?

Acknowledgment

Part B was adapted for use in this exercise with permission from B. S. Meyer, D. B. Anderson, and C. A. Swanson. 1955. *Laboratory Plant Physiology*, 3rd edition. D. Van Nostrand Co., Inc., Princeton, N. J., p. 18.

References

Avers, C. 1982. *Basic Cell Biology*, 2nd edition. Boston: Willard Grant Press.

Giese, A. C. 1979. *Cell Physiology*, 5th edition. Philadelphia: W. B. Saunders Co.

Loewy, A. G. and P. Siekevitz. 1970. *Cell Structure and Function*, 2nd edition. New York: Holt, Rinehart and Winston, Inc.

Raven, P. H., R. F. Evert, and H. Curtis. 1981. *Biology of Plants*, 3rd edition. New York: Worth Publishers, Inc.

Membranes and permeability

Introduction

Membranes that are described as being differentially permeable regulate the passage of diverse materials into and out of the cell, organelles, and vacuoles at different rates. Water seems to pass through membranes exceedingly fast, governed by the osmotic conditions from one cell to the next. Other materials, however, particularly solutes, may move through membranes so slowly that sometimes they seem to be excluded. The idea that all solutes move very slowly through membranes can be somewhat misleading, especially in view of the fact that solute transport is a vital function of living cells. Solute movement through membranes is regulated by mechanisms such as changes in membrane permeability, membrane structure, carrier systems, and environmental conditions.

The exact structure of membranes is unknown, although there is ample evidence that membranes are composed of proteins, lipids, and carbohydrates. The structural changes and arrangement of the membrane components play an important role in membrane permeability. The following exercise is designed to illustrate some factors that influence the permeability of membranes.

Purpose

To study the permeability of living tissues to acids and bases, various ions, and organic substances; to observe factors that influence the properties and permeability of membranes.

Materials

Rhoeo leaves	20 ml glycerol (1M)
beet root (flower petals, red cabbage, or leaves from *Fuchsia* containing anthocyanins)	20 ml glucose (1M)
	20 ml tertiary butyl alcohol (1M)
Tradescantia or red onion leaves	20 ml ethylene glycol (1M)
10 ml KOH (0.025N)	20 ml methanol (1M)
10 ml NH_4OH (0.025N)	2 test tubes
10 ml HCl (0.025N)	5 watchglasses
10 ml acetic acid (0.025N)	5 beakers (80 ml)
20 ml acetone (80%, v/v)	microscope slides and coverslips
50 ml NaCl (4%, w/v)	microscope
50 ml $CaCl_2$, (0.2%, w/v)	hand lens

Procedure

A. Permeability of living tissues to acids and bases

Prepare a number of strips of the lower epidermis of *Rhoeo* (various flower or other tissues containing anthocyanins may be used) and float them on distilled water. Into separate watchglasses or dishes, pour a small portion of the following:

Distilled water

HCl (0.025N)

Acetic acid (0.025N)

KOH (0.025N)

NH_4OH (0.025N)

Place two strips of the epidermal tissue into distilled water, two in KOH, and six into the NH_4OH solution. Record the time required for blue coloration to develop in the strips after immersion.

After the strips in the NH_4OH solution have turned blue (entirely or in part), transfer four of the tissue pieces to a beaker of distilled water. From the beaker of distilled water, transfer two of the pieces to the acetic acid solution and two to the HCl. Record the time required for a color change to take place.

When the color change is completed, transfer the strips from the acid solution into water and then back to the NH_4OH. Again, record the time required to bring about a color change.

Repeat the above procedures several times and determine an average time required for color change in the tissues after several transfers to acid and vice versa.

Observations:

B. **Effect of ions, acetone, and extreme temperatures on properties of beet root cell membranes**

Cut ten slices (twenty slices for duplicate treatments) of beet root, each measuring approximately 1 cm thick, 2.5 cm long, and 2 cm wide. As the slices are cut, place them into a beaker of distilled water and let them remain there for about five minutes. Then decant the water from the beaker and add enough distilled water to cover the slices. Repeat this procedure until the water covering the slices appears to be free of red pigment released from the injured cells.

Transfer a washed beet root slice into two separate test tubes labeled No. 5 and No. 6, respectively. Stopper the test tubes (a cotton plug should be sufficient) and place test tube No. 5 in a freezing compartment and test tube No. 6 in a refrigerator (4°C) for thirty to sixty minutes or until the beet root slice in the freezer looks frozen. After incubation of the slices, set up the following treatments in test tubes with one beet root slice per test tube:

Tube 1: beet root slice plus 20 ml distilled water

Tube 2: beet root slice plus 20 ml 4% NaCl

Tube 3: beet root slice plus 10 ml 0.2% $CaCl_2$, plus 10 ml distilled water

Tube 4: beet root slice plus 10 ml 0.2% $CaCl_2$, plus 10 ml 4% NaCl

Tube 5: beet root slice plus 20 ml 80% (v/v) acetone–water

Tube 6: beet root slice (frozen) plus 20 ml distilled water

Tube 7: beet root slice (4°C) plus 20 ml distilled water

Tube 8: beet root slice plus 20 ml distilled water

Tube 9: beet root slice plus 20 ml distilled water; tube placed in boiling water bath

Allow the above treatments to stand for forty-five minutes and then decant the fluid from each tube. Make observations and comparisons of the solutions; or, for more precise observations, measure the absorption of each solution at 475 nm with a colorimeter or spectrophotometer. Record the results.

Observations and results:

C. Membrane permeability to organic substances

Record the time necessary to plasmolyze epidermal cells of *Tradescantia*, red onion leaves, or cells of similar tissue in a few drops of 1M glycerol solution on a glass slide.

After making a microscopical examination to determine that plasmolysis has occurred, allow the tissue to remain in the glycerol until the cells have recovered from the plasmolyzed condition. Record the time necessary for this recovery.

Now add several drops of distilled water to the tissue to wash off the glycerol. After adding the water, quickly examine the cells under the microscope. What happens?

Follow the above procedure using the following organic substances (1M): glucose, tertiary butyl alcohol, ethylene glycol, and methanol. Remember to record the time necessary for the cells to deplasmolyze in each of the solutions. The relative times for recovery of the cells from the plasmolyzed condition will provide a rough indication as to the relative order of penetration of the substances into the cells. Also indicate for each compound the molecular weight and solubility in water.

Observations and time recorded:

Results/ Conclusions

For each section of the procedure, design and present an appropriate table (or list of observations) to illustrate the results. Interpret the results according to the following guidelines:

1. What is the explanation for plant cells not normally rupturing when placed in pure water?

2. What is the effect of acid and base on the color of the anthocyanins?

3. What are anthocyanins (chemically) and where are they located in the cell?

4. Explain any differences observed with respect to the bases and acids used.

5. Present any reasons that you might think of that would explain the effects of certain ions on the properties of cell membranes with respect to altering their permeability.

6.   Why did cells deplasmolyzed in glycerol suddenly burst after being placed in pure water?

7.   What chemical and physical features of organic compounds facilitate their penetration into plant cells?

References

Brockerhoff, H. 1977. Molecular designs of membrane lipids. In Van Tamelen, E. E. (ed.), *Macro and Multimolecular Systems*. New York: Academic Press, pp. 1–20.

Clarkson, D. T. 1984. Ionic relations. In Wilkins, M. B. (ed.), *Advanced Plant Physiology*. London: Pitman, Publishing, Limited, pp. 319–353.

Davson, H. and J. F. Danielli. 1952. *The Permeability of Natural Membranes*, 2nd edition. New York: Macmillan Co.

Giese, A. C. 1979. *Cell Physiology*. Philadelphia: W. B. Saunders Co.

Hall, J. L. and D. A. Baker. 1978. *Cell Membranes and Ion Transport*. New York: Longman, Limited.

Os Den Kamp, J. A. F. 1979. Lipid asymmetry in membranes. *Ann. Rev. Biochem*. 48: 47–71.

Shinitzky, M. and P. Henkart. 1979. Fluidity of cell membranes — current concepts and trends. *Int. Rev. Cytol*. 60: 121–147.

Singer, S. J. 1974. The molecular organization of membranes. *Ann. Rev. Biochem*. 43: 805–833.

Singer, S. J. and G. L. Nicolson. 1972. The fluid mosaic model of the structure of cell membranes. *Science* 175:720–731.

Stadelmann, E. J. 1969. Permeability of the plant cell. *Ann. Rev. Plant Physiol*. 20: 585–606.

EXERCISE 4

Osmotic potential

Osmosis may be thought of as a special type of diffusion, the movement of water through a differentially permeable membrane. Although in non-biological systems, solvents other than water can be included in the definition, our primary concern with this process in plants is the diffusion of water.

In plant cells, the osmotic potential is a measure of the absence of energy or capacity of the water of a solution (as compared with pure water) to flow through a membrane. Thus, plant scientists use the term *osmotic potential*, designated ψ_s, to describe the absence of energy in a solution due to the number of solvent–solute interactions, as compared with pure water under ideal osmotic conditions.

In considering Gibbs free energy relationships, a negative sign for osmotic potential value is justified because the solvation process is characterized by:

$$G_2 - G_1 = \Delta G$$

where:

G_2 = situation after the solute is dissolved

G_1 = situation before the solute is dissolved

During the preparation of a solution, there is a net mean loss of translational kinetic energy necessary for osmosis because of the work being accomplished through solvent–solute molecular interactions.

The importance of the osmotic potential is that it characterizes a solution in several ways: It indicates the maximum pressure (osmotic pressure) that might develop if the solution were allowed to come to equilibrium with pure water in an ideal osmotic system; it is proportionately related to the amount of solute in a solution (that is, it becomes proportionately more negative as solute is dissolved); and it is therefore proportional to the decrease in capacity of water flow (due to the degree of solvent–solute interactions).

The following exercise is designed to illustrate the classical plasmolytic method for determining the osmotic potential of plant cells. This method involves the use of a graded series of solutions of varying concentrations of solute representing a range of osmotic potentials. Since the solutions are not under pressure, the osmotic potential of each is equal to the water potential. Further, the solutions are prepared so that some of them are hypotonic and others are hypertonic to the cells to be treated. Strips of plant tissue, preferably containing anthocyanins, are incubated in the solutions for a prescribed period of time, removed, and examined under the microscope.

Examination of the strips incubated in the different solutions will show tissue in which all the cells are turgid (that is, the protoplasts are pushed against the cell wall), tissue in which most of the cells are plasmolyzed (that is, protoplasts are pulled away from the cell wall), and tissue in which 50% of the cells are plasmolyzed (incipient plasmolysis). In cells containing anthocyanins, the condition of the cells are easily observed. For example, in turgid cells, the anthocyanins appear to be evenly distributed throughout the cells and adjacent to the cell walls; in plasmolyzed cells, the anthocyanins are concentrated in only part of the cell and away from the wall. Where the cells of a tissue strip showing the plasmolyzed condition number approximately 50%, these are considered to be at incipient plasmolysis. At incipient plasmolysis, the turgor pressure (pressure potential) of the cell is zero and the osmotic potential of the cell contents is equal to the water potential of the cell and also equal to the water and osmotic potential of the external solution. Knowing the osmotic potential value of the external solution causing the incipient plasmolysis, we may determine the average osmotic potential of the cells of the tissue.

Purpose

To illustrate the classical plasmolytic method for determining the osmotic potential of cell solutions.

Materials

plant material such as red onion or any of the following: *Rhoeo discolor, Zebrina pendula* (lower epidermis), *Elodea canadensis* (*Elodea*), whole leaf, various filamentous algae (*Spirogyra*, etc).

separate solutions (50 ml each) of sucrose and NaCl of the following molar concentrations: 0.30,

0.25, 0.24, 0.22, 0.20, 0.18, 0.16, 0.14, 0.12, 0.10

watchglasses, spot plate, or small beakers

microscope

slides and coverslips

small droppers and bottles

razor blade

forceps

Procedure

Cut fresh sections of the lower epidermis from the midrib of a colored *Zebrina* leaf or red onion. Quickly drop the cut strips into a 0.30M sucrose solution contained in a watchglass. After five to ten minutes, mount several of the sections on a slide in a drop of the same solution. Note if all the cells are plasmolyzed. If so, cut fresh sections and place them in a 0.25M sucrose solution. If this concentration causes plasmolysis of all the cells, try a 0.24M sucrose solution and so on until it can be demonstrated that one sucrose concentration plasmolyzes the majority of the cells while in the next lower sucrose solution about 50% of the cells show signs of plasmolysis (incipient plasmolysis). Record the results in Table 4–1.

TABLE 4–1   Results of osmotic potential determinations

Sucrose Solutions at 20°C		Relative Degree of Plasmolysis
Molarity	Osmotic Potential (Atm)	
0.30	– 8.1	
0.25	– 6.7	
0.24	– 6.4	
0.22	– 5.9	
0.20	– 5.3	
0.18	– 4.7	
0.16	– 4.2	
0.14	– 3.7	
0.12	– 3.2	
0.10	– 2.6	

Before reaching a decision about the limiting concentration, be certain that sufficient time has been allowed for plasmolysis to occur. When incipient plasmolysis is observed in the section, repeat the test several times with fresh solution and tissue pieces until you are sure that a given sucrose solution produces incipient plasmolysis. In this condition, the osmotic potential of the cell solution is equal to the osmotic potential of the external solution.

Follow the same procedure outlined above, but substitute a salt solution (NaCl) of the same molarity as that sucrose solution which produced incipient plasmolysis. If the cells are all plasmolyzed, determine the concentration of salt (molar basis) necessary for incipient plasmolysis of the cells. Indicate the salt solution that produces incipient plasmolysis of the tissue. Also show your calculations for determining the osmotic potential for the salt solution that produces incipient plasmolysis in the tissue studied.

Observations:

Results/
Conclusions

Present your observations concerning the condition of the tissues in the various sugar solutions tested. Table 4–1 gives the osmotic potential of molar sucrose concentrations at 20°C. Assuming that the pressure increases $1/273$ for each degree above 20°C, and that there is a direct proportionality between concentration and osmotic potential of a solution, determine the osmotic potential of the cell sap of the *Zebrina* leaf.

Indicate in your results the effect of various sodium chloride concentrations on plasmolysis. When dealing with an electrolyte, what additional factor would have to be considered in the determination of osmotic potential?

In concluding remarks, explain why the degree of plasmolysis caused by solutions of low solute concentration is not the same as that caused by solutions of high solute concentration.

Present the equation for determining the osmotic potential of an ideal nonelectrolyte solution and the adjustment necessary for electrolytes. What does the cryoscopic method for determining osmotic potential entail? Do certain plants have any ecological advantages by possessing cells with high osmotic potentials?

References

Baker, D. A. 1984. Water relations. In Wilkins, M. B. (ed.), *Advanced Plant Physiology*. London: Pitman Publishing, Limited, pp. 297–318.

Boyer, J. S. 1969. Measurement of the water status of plants. *Ann. Rev. Plant Physiol.* 20: 351–364.

Dainty, J. 1976. Water relations in plant cells. In Lüttge, U. and M. G. Pitman (eds.), *Encyclopedia of Plant Physiology*, New Series, Vol. 2, Transport in Plants, Part A: Cells. Berlin: Springer-Verlag, p. 12.

Kramer, P. J. 1983. *Water Relations of Plants*. New York: Academic Press.
Meidner, H. and D. W. Sheriff. 1976. *Water and Plants*. London: Blackie.

Milburn, J. 1979. *Water Flow in Plants*. New York: Longman Group, Limited.

Nobel, P. 1983. *Biophysical Plant Physiology and Ecology*. San Francisco: W. H. Freeman and Co.

EXERCISE 5

Water potential

In considering plant–water relations, the term *water potential* (ψ_w) is used to express the difference between the energy potential of water (translational kinetic energy of the water molecules, capacity to perform work, or to flow in ideal osmotic systems, and so on) at any point in a system and that of pure water under standard conditions.

The following formula helps clarify the meaning of water potential:

$$\psi_w = \mu_w - \mu_w^\circ = RT \ln \frac{e}{e^\circ}$$

where:

ψ_w = water potential

μ_w = potential of water at any point in a system

μ_w° = chemical potential of pure water

R = gas constant (erg/mole/degree)

T = absolute temperature (°K)

e = vapor pressure of the solution in the system at temperature T

e° = vapor pressure of pure water at the same temperature

The expression $RT \ln (e/e^\circ)$ is zero, meaning that pure water has a potential of zero. In biological systems, however, (e/e°) is generally less than zero, making $\ln (e/e^\circ)$ a negative number. As a result, the water potential of the cell sap is usually negative.

If we dissolve a substance such as sugar in pure water contained in a beaker, the resulting solution has an osmotic potential lower (more negative) than that of pure water. In this situation, we might say that energy is expended in maintaining the solute in solution with a net mean loss in the translational kinetic energy of the water molecules or a decrease in the mean number of water molecules that will flow (as compared to pure water). In the beaker, the solution is not under any pressure. Hence, the turgor pressure is zero, and $\psi_w = \psi_s$. What is important is that an increase in solute will produce a more negative osmotic potential and hence water potential. If the solutions were confined in a cell or similar system to allow for the establishment of turgor pressure (or pressure potential), then the amount of turgor pressure (ψ_p) generated would offset the effect of solute and make the water potential less negative than that of the osmotic potential. The relationship of the water potential (ψ_w) to osmotic potential (ψ_s) and pressure potential (ψ_p) is therefore expressed for plant cells as:

$$\psi_w = \psi_s + \psi_p$$

Thus, the water potential is a function of the osmotic potential, pressure potential (turgor pressure), and any environmental condition that influences these osmotic quantities.

In the following exercise, we will illustrate the gravimetric method for determining the water potential of plant tissue. (A second method, Chardokov's, or the "falling drop" method, will be the subject of Exercise 6.) The gravimetric method involves the placement of pre-weighed plant tissue (potato tuber cylinders) into a graded series of sucrose solutions of known concentrations and osmotic potentials. The osmotic potential of the solutions in a beaker is equal to the water potential.

Cylinders of potato tuber tissue are weighed and then placed into the solution (one cylinder per solution) and incubated for a predetermined time. They are then removed and weighed. The weight gain or loss is plotted against the osmotic potential or water potential of each solution ($\psi_s = \psi_w$). When the points are connected, the intercept at the abscissa (though zero) represents the water potential of the tuber cells. The water potential of the solution corresponding to the intercept point is equal to that of the tissue (no weight gain when ψ_w of the tissue is equal to ψ_w of the solution).

Purpose

To illustrate the classical method for determining the water potential of plant tissue.

Materials

plant material: tubers of white potato or sweet potato, beet roots, fleshy fruit (pears or apples)

cork borer

razor blade

separate sugar solutions (150 ml of each) of the following molarities: 0.15, 0.20, 0.25, 0.30, 0.35, 0.40, 0.45, 0.50, 0.55, 0.60

sets of 250 ml beakers (11 per set)

pan balance

graduated cylinder

paper towels

filter paper

aluminum foil

Procedure   Using a sharp cork borer, punch out twenty-two cylinders (two at a time) of equal length (4.5 cm) and about 1 cm in diameter from one peeled and washed potato (no suberized peripheral layers) or from one of the other plant sources indicated above. Immediately after obtaining a cylinder, trim the ends so that the cylinder is 4 cm long. Blot lightly between two pieces of filter paper and temporarily store the cylinders in a closed moist chamber lined with moistened paper towels.

After obtaining the desired number of cylinders, weigh pairs of cylinders as quickly as possible and record the total fresh weight. Immediately after weighing, place each pair in a beaker containing 150 ml of one of the solutions indicated in Table 5–1.

TABLE 5–1   The water potential of potato tuber tissue

Sucrose Solutions (Molarity)	Osmotic Potential (Atm)	Fresh Weight of Cylinders		Change in Weight*
		Initial	Final	
0.15	− 4.0			
0.20	− 5.3			
0.25	− 6.7			
0.30	− 8.1			
0.35	− 9.6			
0.40	− 11.1			
0.45	− 12.7			
0.50	− 14.3			
0.55	− 16.0			
0.60	− 17.0			

* Before the value, indicate weight gain (+), weight loss (−), or no gain or loss (±).

Repeat the procedure until a pair of cylinders is immersed in each of the different sugar solutions. Remember to record the initial total fresh weight of each pair of cylinders before immersing them in the appropriate sugar solution.

Cover the beakers with aluminum foil and set aside for six hours at room temperature. If necessary, place the beakers in a refrigerator for twenty-four hours before making final weight determinations. Alternatively, each of the cylinders can be cut into four pieces before weighing. This will enable equilibrium to be reached during a three-hour laboratory period.

After the prescribed time, remove the cylinders one pair at a time from the solutions. Quickly blot the cylinders with filter paper and measure the final total fresh weight. For each pair of cylinders, calculate the increase or decrease in the original weight and enter the results in Table 5–1.

Calculations:

Results/
Conclusions

Using the graph provided (Figure 5–1), plot the change in weight (ordinate) against the appropriate concentration of sucrose (abscissa).

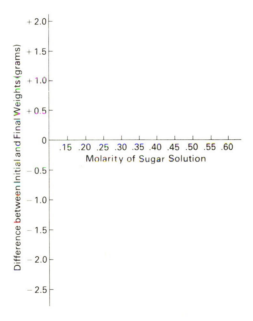

FIGURE 5–1   Weight change in cylinders of potato tuber tissue as a function of the molarity of the ambient sugar solution.

The water potential of the tissue is equal to the water potential of that sugar solution which does not cause an increase or decrease in cylinder weight. Therefore, the point of intersection of the plotted line with the abscissa represents the water potential of the tissue. Explain the rationale behind this statement.

Indicate the water potential of the potato tissue at the beginning of the experiment and the water potential of the cylinders in each solution at equilibrium.

In concluding statements, account for the differences in the amount of water taken up or lost by the tissue. In addition, indicate the relationship of the water potential, osmotic potential, and turgor pressure of the cells and of the various sugar solutions contained in the beakers. You may use equations to answer questions. Of the three osmotic quantities, which one seems to be the most important in regulating the movement of water in plants? Why?

Your interpretation of the results should also take into account some of the technical problems inherent in the procedure for obtaining accurate and predictable values.

References

Boyer, J. S. 1969. Measurement of the water status of plants. *Ann. Rev. Plant Physiol.* 20: 351–364.

Kozlowski, T. T. (ed.). 1968–1978. *Water Deficits and Plant Growth*, Vols. 1–5. New York: Academic Press.

Kramer, P. J. 1983. *Water Relations of Plants*. New York: Academic Press.

Kramer, P. J., E. B. Knipling, and L. N. Miller. 1966. Terminology of cell-water relations. *Science* 153: 889–890.

Meyer, B. S. and A. M. Wallace. 1941. A comparison of two methods of determining the diffusion pressure deficit of potato tuber tissue. *Amer. J. Bot.* 28: 838–843.

Milburn, J. 1979. *Water Flow in Plants*. New York: Longman Group, Limited.

Sutcliffe, J. 1968. *Plants and Water*. New York: St. Martin's Press.

Measurement of the water potential of plant tissues by the falling drop method

Introduction

The "falling drop" or Chardokov's method for determining water potential in plant tissue is often modified, depending upon the kind of experiment to be performed. Normally, the method involves the establishment of a duplicate series of sucrose solutions in test tubes. These solutions range from 0.15 to 0.50 with increments of 0.5 molality. Plant tissue pieces of similar sizes and weight are added to the solutions in each test tube of one series. To the other series, only a drop of methylene blue is added to each solution.

After the tissue pieces have incubated for a prescribed time, they are removed from the test tubes. A drop from the corresponding solution of the same molality containing the methylene blue is introduced under the surface of the solution in which a tissue was incubated. The direction of the drop movement will indicate whether the incubating solution gained water from the tissue or lost water to the tissue. If the drop diffuses uniformly into the solution, the water potential of the solution is equal to that of the tissue because there has been no net gain or loss of water. The procedure to illustrate the falling drop method in the following exercise has been modified to facilitate observations by the class during the time allotted.

Purpose

To determine the water potential of potato tuber tissue by the falling drop method.

Materials

peeled potato (turnips, carrots, or leaf tissue)	8 capillary pipets (or medicine droppers drawn to a fine tip)
60 ml sucrose solution (0.5M)	pipet (10 ml graduated or 10 ml dispensing burette)
10 ml methylene blue solution (0.2% w/v in water)	test tube racks (to hold 16 test tubes)
8 test tubes (15 × 150 mm, preferably graduated at 10 ml levels)	cork borer (4 mm internal diameter)
8 test tubes (13 × 100 mm)	forceps

Procedure

A. Control and test series of different sucrose concentrations

Support eight test tubes (15 × 150 mm) in a row and add the required volumes of a 0.5M sucrose solution which, when diluted to 10 ml with water, will give a range of molarities extending from 0.15M to 0.50M in steps of 0.05M. Label this series of tubes the *control series*. Then set up another series of the same sucrose concentrations (labeled *test series*) by transferring 4 ml of each solution of the control series to smaller test tubes (13 × 100 mm). Place the control series of solutions as a row in order of increasing concentration. Then place the test series of tubes in corresponding positions in the second row. Stopper the tubes until ready for use.

B. Determination of water potential of potato cylinders

Note: The following procedure differs slightly from standard techniques so that water potential determinations can be studied with time.

With a cork borer (internal diameter of 4 mm), cut a cylinder from a freshly peeled potato tuber and trim to 2 cm in length. Immediately transfer the cylinder after cutting to the first sucrose solution of the test series in the rear row. Repeat the procedure for the remaining seven tubes of the test series. The test solutions should cover the plant material.

When each tube of the test series contains a tissue cylinder, add a drop of methylene blue solution (0.2% w/v in water) to each test solution and shake.

After thirty minutes, withdraw a small amount of the test solution from the first tube of the test series with a capillary pipet. Then hold the pipet about an inch under the surface of the solution in the control series, which initially contained the same concentration of sucrose. Slowly release a drop of the blue test solution, and against a white background determine whether the drop rises, falls, or simply diffuses out. If the drop rises, the test solution has decreased in density over its initial value. Conversely, if the density has increased, the drop of test solution will fall. Accordingly, if the density of test solution is the same or close to the control solution, the drop will not move up or down appreciably, but will diffuse out. However, since a closely graded concentration series was not used, the actual "null point" will have to be interpolated.

Perform the basic procedure on all the test solutions and repeat at thirty-minute intervals during the next two hours. Between measurements, keep each pipet in its respective test solution and avoid transferring significant quantities of the control solution to the test solutions and vice versa. Also, before withdrawing a test drop, flush the pipet out with test solution several times. Record the results in Table 6–1.

TABLE 6–1   Results of determinations of the water potential of potato tuber tissue

Control Tube (Number)	Sucrose Solutions (Molarity)	Osmotic Potential (Atm at 20°C)	Direction of Drop Movement with Time (Min.) *						
			30	45	60	75	90	105	120
1	0.15	– 4.0							
2	0.20	– 5.3							
3	0.25	– 6.7							
4	0.30	– 8.1							
5	0.35	– 9.6							
6	0.40	– 11.1							
7	0.45	– 12.7							
8	0.50	– 14.3							

* Indicate the relative movement of the test drops in the control series by: up, down, or null. If no sharp null point is observed, it should be interpolated.

Results/ Conclusions

Interpret your results as recorded in Table 6–1 and present an approximate water potential value for the tissue tested. Remember that the water potential of the tissue is equal to the osmotic potential of that sugar solution in which there is no net movement of water into or out of the tissue.

In concluding remarks, account for the rise or fall of the test drops when placed into the control solutions. Under what conditions would the densities of the various test solutions/increase, decrease, or stay the same? How does the water potential value of a tissue affect the changes in density of the surrounding solution?

Although the water potential value of a tissue can be determined after a brief period of immersion, it should theoretically be the same as after a longer time. However, in practice this value slowly shifts with time. How do you account for the deviation from theory?

If Exercise 6 was not performed, indicate the relationship of the water potential to osmotic potential and turgor pressure of the cells and of the various sugar solutions in the tubes. Of the three osmotic quantities, which one seems to be the most important in regulating the movement of water in plants? Why? In addition, how might these osmotic quantities be explained in terms of the free energy concept for water movement in defined osmotic systems?

Acknowledgment

This exercise was taken from the work of E. B. Knipling and was suggested for class use by B. R. Roberts, United States Department of Agriculture, Shade Tree Research Laboratory, Delaware, Ohio.

References

Boyer, J. S. 1966. Isopiestic technique: Measurement of accurate leaf water potentials. *Science* 154: 1459–1460.

Brix, H. 1966. Errors in measurement of leaf water potential of some woody plants with the Schardakow dye method. *Can. Dept. Foresty Publ.* 1164.

Goode, J. E. And T. W. Hegarty. 1965. Measurement of water potential of leaves by methods involving immersion in sucrose solutions. *Nature* 206: 109–110.

Knipling, E. B. 1967. Measurement of leaf water potential by the dye method. *Ecol.* 48: 1038–1041

Rehder, H. and K. Kreeb. 1961. Vergleichende Untersuchungen zur Bestimmung der Blattsaugspannung mit der gravimetrischen Methode und der Schardakow-Methode. *Ber. Dutsch Bot. Ges.* 74: 95–98.

Shardakov, V. S. 1948. New field method for the determination of the suction pressure of plants. *Dokl. Akad. Nauk. SSSR.* 60: 169–172.

(Also see references following Exercises 4 and 5.)

EXERCISE 7

The cohesion of water and lifting power of evaporation

Introduction

To illustrate the lifting power of evaporation, a hollow glass tube is filled with water and immersed at one end into a stoppered bottle containing mercury covered with water. The tube is filled with water so that there is an unbroken connection between the mercury in the bottle and the water in the tube. The other end of the tube is then introduced into a stoppered porous vessel filled with water so that a connection is made between the water held in the "evaporating" vessel and the water in the tube. This results in an unbroken column of water throughout the system. When water evaporates from the vessel at the top of the system, water and then the mercury will move up the tube. The movement of water and mercury up the tube is due to the driving force of evaporation (less negative chemical potential at the surface of the porous vessel to very negative chemical potential of the atmosphere) and cohesive strength of the liquid molecules. The column of liquid will break when its cohesive strength is overcome by gravitational pull or by an interruption of the column by air. In this exercise, you will appreciate the lifting power of evaporation when you observe the height to which the mercury is pulled in the column, for liquid mercury is approximately fourteen times heavier than water.

Purpose

To study a physical system used to explain the movement of water in plants based on evaporation and the cohesion of water.

Materials

porous clay cup (see Appendix II for source)	two-hole rubber stopper
glass tube (J-shaped, 100 cm long, with 1 mm bore; see Figure 7-1)	mercury (clean and dust-free)
	beaker (1 liter)
one-hole rubber stopper (to fit clay cup and to accommodate one end of the J-shaped glass tube)	ring stand
	hot plate
	supports for beaker and hot plate
glass tubing (for side arm)	gelatin solution (20% w/v, if necessary)
wide-mouthed bottle (see Figure 7-1)	

Procedure

A. Assembling the apparatus

All parts should be thoroughly cleaned with boiling cleaning solution and rinsed a number of times with distilled water before the apparatus is assembled.

In assembling the apparatus (see Figure 7-1), the vertical part of the 1 mm bore glass tube should be 100 cm or more in height, and its lower end should be immersed in mercury to a depth of 1 cm. The mercury used in the reservoir should be perfectly clean and free from dust or other contaminants.

The porcelain cylinder upon which the success of this exercise depends should be as fine-pored as possible and free from cracks or imperfections in the form of large pores. However, even somewhat imperfect clay evaporating surfaces can be modified to give satisfactory results (discussed below).

After the apparatus has been assembled and filled with recently boiled distilled water, immerse the porous cylinder in a beaker of distilled water. To ensure that the system is completely free from undissolved air bubbles, heat the water in the beaker to just under boiling and maintain the heat for two to three hours while hot water is allowed to syphon through the apparatus and out of the side arm.

After the prescribed time, remove the source of heat and allow the entire apparatus to cool to room temperature. During the cooling process, water will continue to syphon slowly through the apparatus.

B. Lifting power of evaporation and cohesion of water

After the system has cooled, simply remove the beaker of water in which the cylinder is immersed. If there is any doubt concerning the perfection of the porcelain cylinder, pour a hot 20% (w/v) gelatin sol over the surface of the cylinder as it is removed from the beaker of water. This gelatin coating forms the evaporating surface in place of the porous clay.

Observe the water rising up the vertical glass tube, followed by the rising column of mercury. The rate of ascent of liquid can be accelerated by allowing a gentle breeze from an electric fan to play on the cylinder.

FIGURE 7–1   Apparatus for demonstrating the cohesion of water and lifting power of evaporation. (Used with permission B. S. Meyer, D. B. Anderson, and C. A. Swanson. 1955. *Laboratory Plant Physiology*, 3rd edition. D. Van Nostrand Co., Inc., Princeton, N.J., p. 54.)

Results/ Conclusions

Present and interpret your observations on the basis of the answers to the following questions:

1.   Explain why the water and mercury rise, and what determines the rate of rise.

2.   How do you account for the rise of the mercury above the level that can be accounted for by atmospheric pressure?

3.   How does this experiment demonstrate the existence of a cohesion force in water, and why does the water not pull away from the mercury when tension develops in the column of water?

4.   How do the principles illustrated in this exercise apply to the theories regarding the rise of water through the stems of plants?

Acknowledgment   This exercise was adapted from B. S. Meyer, D. B. Anderson, and C. A. Swanson, 1955. *Laboratory Plant Physiology,* 3rd edition. D. Van Nostrand and Co., Inc., Princeton, N.J., pp. 54–56.

References   Askenasy, E. 1896a. Über des Saftsteigen. *Verhandl. naturhist. med. Ver. (Heidelberg) N.S.* 5: 325–345.

Askenasy, E. 1896b. Beiträge zur Erhlärung des Saftsteizens. *Verhandl. naturhist. med. Ver. (Heidelberg) N.S.* 5: 429–448.

Dixon, H. H. 1914. *Transpiration and the Ascent of Sap in Plants.* London: Macmillan Co.

Hall, A. E. and M. R. Kaufmann. 1975. Regulation of water transport in the soil–plant–atmosphere continuum. In Gates, D. M. and R. B. Schmerl (eds.), *Perspective of Biophysical Ecology.* Berlin: Springer-Verlag.

Honert, T. H. van den. 1948. Water transport in plants as a catenary process. *Discuss. Faraday Soc.* 3: 146–153.

Zimmermann, U. and E. Steudle. 1978. Physical aspects of water relations of plant cells. *Adv. Bot. Res.* 6: 45–117.

EXERCISE 8

Transpirational pull and root pressure

The cohesion–tension theory of Dixon is the most plausible explanation for the translocation of water in plants. According to this theory, a column of water may be "pulled" upward because of its cohesive and adhesive properties. As water evaporates from the mesophyll cells of a leaf, the water potentials of the mesophyll cells in direct contact with the air spaces of the leaf become more negative. The water lost by the cell surfaces is replaced from the internal water solutions of the cells. The water potentials of the cells adjacent to the air spaces decrease, causing water to move from cell to cell along water potential gradients in an energetically favorable direction (from less negative to more negative water potential). This movement within the leaf is transmitted to the water in the xylem cells in the veins, exerting a pull or creating a state of tension. This tension continues through unbroken columns of water due to the cohesion of water molecules throughout the xylem from the leaves to the roots. As water moves from the roots, the water potentials in the living cells and along the cell walls becomes more negative in relation to the water potential of the soil, thus promoting absorption.

The following exercise is designed to illustrate the lifting power of transpiration by demonstrating the rise of mercury in a closed system due to the influence of a transpiring plant. The phenomenon of root pressure will also be demonstrated. If a well-watered plant is detopped, and the stump attached with a rubber sleeve to a glass tube, we may observe xylem sap exuding from the cut end of the stump. Root pressure may be explained as a pressure developing when transpiration is low in the tracheary cells of the xylem as water accumulates due to osmotic mechanisms created in the root cells by increased solutes followed by water absorption. The pressure is then transmitted to the upper portions of the plant as water is pushed upward. Root pressure is a curious phenomenon and may be partly responsible for the translocation of water and dissolved solutes in low-growing plants. However, it is not considered to be the primary means by which water and solute are translocated in plants.

Purpose

To demonstrate transpirational pull by a leafy shoot, and to estimate the magnitude and variations of root pressure.

Materials

potted *Fuchsia* plant (geranium, tobacco, or other suitable species)

rubber tubing (about 4 cm long)

capillary tubing (3 ft. long, internal diameter 1 to 2 mm)

thread

sealing wax

two-hole rubber stopper

small wide-mouthed bottle containing freshly boiled and cooled distilled water

J-shaped tube (long arm at least 80 cm in length, internal diameter 2 to 3 mm)

chromic acid cleaning solution

small beaker of mercury

razor blade

Procedure

Amputate the stem of a well-watered potted *Fuchsia* plant about six inches above the soil. Hold the section of the stem under water when it is cut. If this is not possible, quickly place the cut end of the shoot in water and cut off an additional 2 cm of stem under water. Maintain the upper portion of the shoot with the cut end in water until you are ready to proceed to Part B.

A.   Root pressure (Figure 8–1A)

Roll onto the stump of the plant remaining in the pot a piece of wetted rubber tubing, leaving about 2 cm of the tubing above the cut surface. Now force one end of a capillary tube (3 ft. long with an internal diameter of 1 to 2 mm) down into the rubber tubing until contact is made with the stump. Support the capillary tube with a ring stand and clamp.

Bind the stump and tube connections with strong thread so that the unions will remain liquid-tight under considerable pressure. Do not injure the bark. Close the open end of the capillary tube with a drop of sealing wax and measure the volume of air in the tubing above the cut stem.

As liquid is forced into the capillary, the volume of air will be reduced but the pressure of the gas will proportionately increase. Therefore, by measuring the volume of the enclosed gas at various times, one can estimate the magnitude and variations of root pressure. Obtain frequent readings of the volume of gas over a period of two days and compute the root pressures.

Readings and calculations:

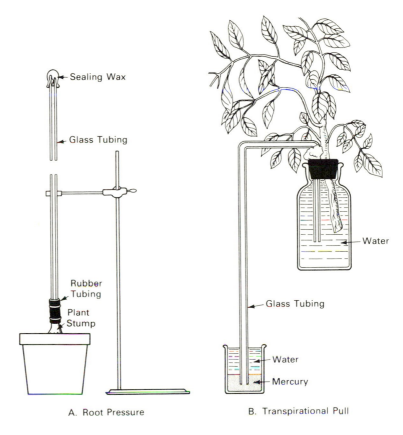

Sealing Wax

Glass Tubing

Rubber Tubing

Plant Stump

Water

Glass Tubing

Water

Mercury

A. Root Pressure

B. Transpirational Pull

FIGURE 8-1   Apparatus for demonstrating transpirational pull and root pressure.

B.   Transpirational pull (Figure 8-1B)

Remove the bark from the base of the plant under water and quickly thrust the xylem through a two-hole rubber stopper so that the stem will be nearly to the bottom of a small wide-mouthed bottle of freshly boiled and cooled water when the stopper is inserted. Quickly insert the stopper (with the shoot) into the bottle.

Prepare a J-shaped glass tube of 2 to 3 mm internal diameter and with a long arm at least 80 cm in length. Thoroughly clean this tube with chromic acid cleaning solution (*Caution!*), fill it with boiled distilled water, and thrust the short arm through the second hole of the rubber stopper nearly to the bottom of the bottle containing the stem.

Immerse the lower end of the long arm in mercury. Eliminate all bubbles and leaks from the system and use paraffin or sealing wax around the stem, capillary tube, and stopper if necessary. Frequently observe and record the maximum height that the mercury level attains.

Observations:

Results/Conclusions

For the root pressure part of this exercise, illustrate the results with a graph in which the root pressures (in atm) are plotted as the ordinate against time (in hours) as the abscissa. With respect to root pressure, base your conclusions on your observations and answers to the following questions:

1. Is root pressure dependent upon living cells in the root?

2. Would a rapidly transpiring plant be expected to exhibit the phenomenon of root pressure?

3. What is Priestley's theory of root pressure?

4. How may root pressure influence sap flow?

5. Is root pressure the predominant force in the translocation of water in plants? Is transpiration?

Present your results concerning transpirational pull as demonstrated, and calculate the height, in feet, to which the shoot should have been able to lift water. Why did the mercury column not rise higher? Under what conditions and how high will atmospheric pressure force a column of mercury? Of water? What conditions in the exercise are probably not attained in the plant?

References

Barrs, H. D. 1966. Root pressure and leaf water potential. *Science* 152: 1266–1268.

Boyer, J. S. 1977. Regulation of water movement in whole plants. *Symp. Soc. Exp. Biol.* 31: 455–470.

Dixon, H. H. 1914. *Transpiration and the Ascent of Sap in Plants.* London: Macmillan Co.

Scholander, P. F., H. T. Hammel, D. Bradstreet, and E. A. Hemmingsen. 1965. Sap pressure in vascular plants. *Science* 148: 339–346.

Thut, H. F. 1928. Demonstration of the lifting power of evaporation. *Ohio J. Sci.* 28: 292–298.

Weatherby, P. E. 1963. The pathway of water movement across the root cortex and leaf mesophyll of transpiring plants. In Rutter, A. J. and F. H. Whitehead (eds.), *The Water Relation of Plants.* London: Blackwell Scientific, p. 86.

Measurement of plant moisture stress, water potential, and xylem tensions with the pressure bomb

Introduction

The pressure bomb is used to determine plant moisture stress and the water potential of leaves on a leafy shoot. It is based on the fact that the water columns in a plant are almost always under tension because of the pull exerted by the osmotic conditions (water potential) of the living cells of the leaves. When the tension in the xylem conducting cells is high, the water potential of the leaf cells is very negative.

If one cuts a leafy stem, the water column is disrupted, and because the water column is under tension it will recede back into the stem toward the leaves. When the leafy end of the shoot is placed into the pressure bomb chamber so that the cut end is protruding out of the chamber and pressure is applied in the chamber, the water column can be forced back to the cut surface. The pressure required to force the water to the surface of the cut can be recorded and is equivalent to the amount of tension (but with opposite sign) that existed in the stem prior to cutting. If low pressure is sufficient to force water to the cut surface, the cells of the leaves have slightly negative water potentials and the shoot was under low moisture stress (tension). If high pressure is required, the water potentials of the leaf cells are very negative and the moisture stress is high. Nevertheless, the pressure required to force water to the cut surface can be determined by reading the gauge on the tank or chamber (depending upon the equipment available) and converted to an equivalent value for a close estimate of the xylem water potential or tension.

In performing the following exercise, be sure to follow the instructions outlined below in addition to those given by the instructor. As a word of caution, be sure not to place any part of your body directly above the pressure chamber opening since the plant material may be extruded through the rubber stopper at high pressures. Also, be sure to exhaust the pressure bomb chamber before attempting to remove the cover.

Purpose

To measure the plant moisture stress (tension) and water potential of a leafy shoot with the pressure bomb apparatus.

Materials

leafy shoots that can be accommodated by the pressure bomb chamber

sharp knife or razor blade

pressure bomb apparatus equipped with portable tank, insertion

tool, and additional inserts (Model 600 may be purchased from PMS Instrument Company, 5945 NW Rosewood Dr., Corvallis, Oregon 97330)

Procedure

A. Preliminary adjustments and precautions

Be sure to check with the instructor concerning any safety precautions and details pertaining to operation of the apparatus. The instructor will provide you with the plant materials to be used and may give a demonstration of the equipment.

B. Preparation of plant material and insertion into pressure bomb chamber

1. Select a leafy shoot for study from different plants or portions of the same plant that can be accommodated by the pressure bomb chamber.

2. Cut the selected shoot (just prior to measurement) with a sharp knife or razor blade. A clean slanting cut surface is best. You may also remove the phloem (bark) up to about one inch away from the cut surface.

3. Place the stem into the rubber stopper insert oriented so that the leafy end will be inside the chamber. Insertion of succulent or fragile tissue is accomplished by using the insertion tool. Place the stopper into the pressure chamber lid and tighten the lid to the chamber.

4. The cut surface of the stem should extend from the chamber about an eighth to a quarter of an inch above the stopper so that sap exudation can be easily observed.

5. Make sure the lid is securely tightened.

C. Application of pressure and readings

1. Observe the cut end of the stem (you may wish to use a hand lens) and slowly introduce the compressed air into the chamber by turning the control valve to Chamber position.

2.   When the sap is first seen at the cut surface (a film of water appears on the cut surface), turn the control valve to the Off position. Read the pressure value on the gauge and record it.

3.   Turn the control valve to the Exhaust position, remove the cover from the chamber, and discard the shoot. You are now ready to begin another measurement.

4.   Repeat the measurements on several shoots for the same or different plants. You may also measure the tension in leaves.

Results/ Conclusions

Record the pressure (in atmospheres) observed to restore the sap to the cut surfaces of the selected shoots or leaf petioles. You should present your results as potentials having opposite signs of the pressures recorded.

For your understanding of this exercise, you should consider the current theories pertaining to the ascent of water in plants, the diurnal fluctuations in xylem tension, and the reasons for the establishment of xylem tension and water stress.

References

Kaufman, M. R. 1968. Evaluation of the pressure chamber technique for estimating plant water potential of forest tree species. *Forest Science* 14: 369–374.

Scholander, P. F., H. T. Hammel, E. D. Bradstreet, and E. A. Hemmingsen. 1965. Sap pressure in vascular plants. *Science* 148: 339–346.

Tyree, M. T. 1976. Negative turgor pressure in plant cells: Fact or fallacy? *Can. J. Bot.* 54: 2738–2746.

Tyree, M. T. and H. T. Hammel. 1972. The measurement of turgor pressure and the water relations of plants by the pressure bomb technique. *J. Exp. Bot.* 24: 267–282.

Waring, R. H. and B. D. Clary. 1967. Plant moisture stress: evaluation by pressure bomb. Science 155: 1248–1254.

Wiebe, H. H., R. W. Brown, T. W. Daniel, and E. Campbell. 1970. Water potential measurements in trees. *BioScience* 20: 225–226.

EXERCISE 10

Stomates

Introduction

With few exceptions (the algae and fungi), stomates (stomata) are widespread in the plant kingdom and are found in great abundance on the epidermal surface of leaves. Stomates are microscopic and are bordered by two guard cells, which control the opening and closing of the stomatal pore. When fully open, the stomatal pore may measure from 2 to 12 μm in width and from 10 to 40 μm in length. Stomata are frequently found in the upper surface of leaves, but in many species they are found on both surfaces. Depending upon the species, the surface of a leaf may contain 1000 to 60,000 stomata per square centimeter. Yet, the stomata are so small that when fully open, they occupy only 1 to 2% of the total leaf surface.

Stomata are physiologically of great significance to plants. It is through the stomatal pore that gas exchange takes place — CO_2 and O_2, important in photosynthesis and respiration, respectively. Also, a significant amount of water absorbed by plants is lost as vapor through open stomata during the course of stomatal transpiration. Certainly, water lost by transpiration is very important in terms of plant growth and survival.

To understand and control stomatal transpiration, plant scientists have expended a good deal of time and resources in studying the anatomy and physiology of stomata. As a result, there have been significant advances in our understanding of the structure of stomata and the mechanism(s) responsible for their opening and closing.

During the following exercises, you will be shown some of the procedures for studying stomata. If possible, you should observe stomata on the surface of leaves of both dicot and monocot plants.

Purpose

To illustrate a silicone rubber technique used in obtaining impressions of the abaxial and adaxial surfaces of leaves; to study the distribution of stomates, shape of the guard cells, and the mechanism of opening and closing of stomates.

Materials

intact plants of bean (*Phaseolus vulgaris* L.), corn (*Zea mays* L.), or other representatives of a dicot and monocot

Tradescantia

potted herbaceous plant (geranium, sunflower, tobacco, etc.)

silicone rubber compound (liquid silicone rubber RTV-21, obtained from Silicone Products Dept., General Electric Co., Waterford, N.Y.)

catalyst (Nuodex or Nuocure 28, obtained from Tenneco Chemicals, Inc., Nuodex Div., Elizabeth, N.J. or Long Beach, Calif.)

silicone aquarium sealant

nail polish (clear or natural)

sucrose solution (1M) in a dropping bottle

benzene

dissecting needle

forceps

microscope slides and coverslips

microscope

light source

Procedure

A.   Silicone rubber and nail polish impressions of leaf surfaces

The procedure for making leaf surface impressions is shown in Figure 10–1. As shown, place about one-half tablespoon of silicone rubber compound in a petri dish and thoroughly mix with it about six drops of the catalyst. The amount of catalyst should be used according to the time desired for hardening of the silicone rubber.

Immediately apply a small amount of the mixture as a thin layer on a portion of the leaf surface to be examined (see Part B). Allow the preparation to harden on the leaf (about thirty to forty-five minutes). After the thin coating has hardened, gently peel it from the leaf surface with a dissecting needle and forceps. The resulting mold is the primary impression and may be stored in an envelope properly labeled as to the genus, species, and the surface (upper or lower) of the leaf. Alternatively, silicon aquarium sealant can be used in the same way. This material needs no external catalyst.

After obtaining the primary impression, coat it with a thin layer of clear or natural nail polish. After the polish hardens, peel the secondary impression from the primary with the aid of a dissecting needle and forceps.

Mount the secondary impression on a microscope slide with the dull side facing upward. Add a drop or two of water, place a coverslip over the slide, and examine with a microscope. For best observation, the mirror and condenser should be adjusted so that the mount is contrasted against its background.

B.   The shape of guard cells and relative distribution of stomates on abaxial and adaxial leaf surfaces

Prepare primary and secondary impressions of the upper and lower leaf surfaces of bean and corn plants growing in the greenhouse. If these plants are not

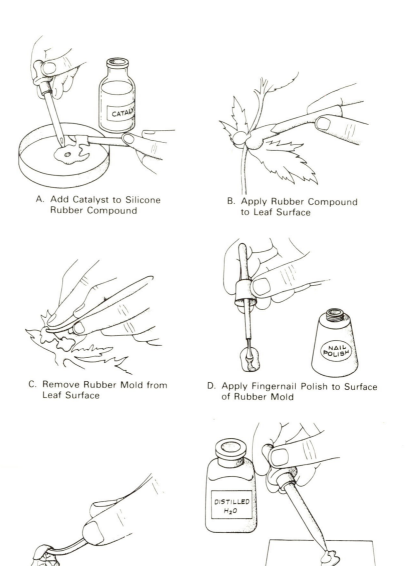

A. Add Catalyst to Silicone
   Rubber Compound

B. Apply Rubber Compound
   to Leaf Surface

C. Remove Rubber Mold from
   Leaf Surface

D. Apply Fingernail Polish to Surface
   of Rubber Mold

E. Remove Fingernail Polish Film
   from Rubber Mold

F. Prepare Replicate as a Wet Mount

FIGURE 10–1   Preparation of leaf impressions.

available, any representative dicot that possesses "kidney bean-shaped" guard cells and a representative grass that possess "dumbbell-shaped" guard cells may be used. It should be emphasized, however, that this silicone rubber method is best suited for obtaining impressions from glabrous leaves.

Study the secondary impressions illustrating the two differently shaped guard cells and make a sketch of your observations. In addition, determine the relative number of stomates on the upper and lower surfaces.

If time permits and at the discretion of the instructor, make five to ten impressions of the lower epidermis of leaves on a plant growing in the greenhouse, and by calculating the area of the high-power field of the microscope, determine the average number of stomates per square centimeter of leaf surface.

Observations and calculations:

C. Opening and closing of stomates

Strip off a small piece of the lower epidermis of a *Tradescantia* leaf that has been kept for some time in bright light. Mount the piece in water on a slide and select under the high-power objective lens the most widely open stomate you can find. Observe which epidermal cells contain plastids. Also estimate the width of the stomatal pore. Now replace the water with a drop of approximately 1M sucrose solution by using a small piece of filter paper to withdraw the water from a side of the cover glass opposite the side where the drop of sucrose solution is applied. Observe the stomate for a few minutes.

After the stomate has closed, replace the sucrose solution with water. Several flushings may be necessary. Stand the microscope in such a position that the slide is exposed to bright light. Observe from time to time and determine the time it takes for the stomate to open again.

Observations:

D. Benzene infiltration method for determining whether stomates are closed or open

Select a potted and well-watered herbaceous plant (geranium, bean, sunflower, tobacco, and so on) that has been exposed to bright light for at least one hour. Place the plant in front of a bright light source and then with a brush apply a small amount of benzene on the lower epidermis of one leaf. Avoid contact with or breathing the benzene. Notice the translucence in the area of the leaf to which the benzene was applied.

Place the plant in darkness for at least one hour and repeat the benzene infiltration test. Record your observations.

Observations:

Results/ Conclusions

Present your results and interpretations for each part of the exercise. In concluding statements, answer the following:

1. What are the advantages and disadvantages of using the silicone rubber impression technique for studying the distribution, size, and shape of the stomatal apparatus?

2. Although not demonstrated in this exercise, do the dumbbell-shaped guard cells characteristic of corn operate differently than the kidney bean-shaped guard cells with respect to the opening and closing of stomates? The latter guard cells possess relatively thickened inner walls, whereas the former do not.

3. On which leaf surface is the majority of stomates found in dicots? In monocots?

4. Does the presence of a large number of small pores contribute to greater water vapor loss than would be expected from fewer but significantly larger pores? Explain.

5. What causes the changes in the osmotic potential of guard cells, and how do changes in the water potential of the guard cells regulate stomatal openings and closings?

6.   What environmental conditions influence the opening and closing of the stomates?

7.   Why will benzene enter a leaf through open stomates when under the same conditions water will not?

8.   Is the benzene method suitable for determining whether or not the stomates are opened or closed?

Acknowledgment

Parts C and D of the exercise were adapted from B. S. Meyer, D. B. Anderson, and C. A. Swanson. 1955. *Laboratory Plant Physiology*, 3rd edition. D. Van Nostrand Co., Inc., Princeton, N.J., p. 46.

References

Alvim, P. de T. and J. R. Havis. 1954. An improved infiltration series for studying stomatal opening as illustrated with coffee. *Plant Physiol.* 29: 97–98.

Barrs. H. D. 1971. Cyclic variations in stomatal aperture, transpiration, and leaf water potential under constant environmental conditions. *Ann. Rev. Plant Physiol.* 22: 223–236.

Boyer, J. S. 1977. Regulation of water movement in whole plants. *Symp. Soc. Exp. Biol.* 34: 455–470.

Cowan, I. R. 1977. Stomatal behavior and environment. *Adv. Bot. Res.* 4: 117–228.

Farquhar, G. D. and T. D. Sharkey. 1982. Stomatal conductance and photosynthesis. *Ann. Rev. Plant Physiol.* 33: 317–345.

Hall, A. E., and E. D. Schulze. 1980. Stomatal response to environment and a possible interrelation between stomatal effects on transpiration and CO_2 assimilation. *Plant Cell Environ.* 3: 467–474.

Horton, R. F. 1971. Stomatal opening: The role of abscisic acid. *Can. J. Bot.* 49: 583–585.

North, C. 1956. A technique for measuring structural features of plant epidermis using cellulose acetate films. *Nature* 178: 1186–1187.

Raschke, K. 1975. Stomatal action. *Ann. Rev. Plant Physiol.* 26: 309–340.

Willmer, C. M. 1983. *Stomata.* London: Longman, Limited.

Zelitch, I. 1963. Stomata and water relations in plants. *Connecticut Agri. Exp. Sta. Bull. No. 664.*

Zelitch, I. 1965. Environmental and biochemical control of stomatal movement in leaves. *Biol. Rev.* 40: 463–482.

Zelitch, I. 1969. Stomatal control. *Ann. Rev. Plant Physiol.* 20: 329–350.

EXERCISE 11

Transpiration rates as determined by loss of weight of potted plants

Perhaps the simplest means to observe transpiration in plants is to weigh a potted plant at the beginning and end of a prescribed period of time. It is important that the soil surface be covered and the pot wrapped with aluminum foil or some other water repellent material in order to retard evaporation from surfaces other than the plant. The loss of weight over a short period of time will be almost entirely due to transpiration.

The method applied in this exercise is restricted to the use of small plants in pots. However, it is suitable for studying the influence of various environmental factors on the relative rates of transpiration and the magnitude of transpiration in small plants.

Purpose

To estimate the transpiration rate by recording the changes in weight of potted plants exposed to different environmental conditions.

Materials

6 potted plants (bean, tobacco, tomato, or sunflower)	paraffin
	dark box or cabinet
oilcloth or aluminum foil	planimeter (not essential)
string or rubber bands	

Procedure

Water six plants excessively that have been grown in four-inch pots and allow the pots to stand until all the excess water has drained off. Then wrap each pot completely with a large piece of oilcloth (aluminum foil may also be used). Also wrap a little cotton around the base of the stem and cover the surface of the soil with two additional pieces of oilcloth that have been slit to a small hole in the center to accommodate the stem of the plant. The oilcloth covering the soil surface should be arranged so that the slits do not coincide and should be held snugly in place with a string or rubber band. Seal the cotton and any openings in the oilcloth with paraffin.

Immediately record the weight of all the plants in their respective pots and subject them to any desired environmental conditions. As a suggestion, the plants may be subjected to the following treatments:

Darkness

Bright light

Subdued light

Subdued light in front of an operating fan

Bright light in front of an operating fan

Light, humidity, and temperature conditions of the greenhouse

Regardless of the environmental conditions, weigh each plant at one- or two-hour intervals over an eight-hour period or longer, if desired. Record the weights and determine the amount of water loss by subtracting each weight obtained from the initial and preceding weight.

Observations:

The plant placed in the greenhouse may be studied for several days and weight determinations made at two-hour intervals every day. The first determination each day should be made as early as possible in the morning, and the final reading made in the evening. Also, if suitable recording in instruments are available, the daily periodicity of of plants may be correlated with those changes in environmental conditions.

At the end of an experiment, remove all the leaves from the plant(s) and make tracings of the leaves on paper. Use a planimeter, and from the tracings determine the total leaf area of the plant in square decimeters. Calculate the amount of water loss per square decimeter of leaf area for each two-hour period.

Calculations:

Results/
Conclusions

Illustrate the results with a graph in which the water loss per square decimeter of leaf area is plotted as the ordinate against time as the abscissa.

In concluding statements, interpret your results with respect to the principal environmental factors that influence transpiration rates and to the roles played by the diffusive capacity of the stomates. Would you expect plants growing in the field to exhibit the same general type of transpiration periodicity as those growing in pots? What are some of the disadvantages that might be involved in using the method outlined in this exercise for determining transpiration rates over a long period of time?

Acknowledgment

The general techniques outlined in this exercise were adapted from B. S. Meyer, D. B. Anderson, and C. A. Swanson. 1955. *Laboratory Plant Physiology,* 3rd edition. D. Van Nostrand Co., Inc., Princeton, N.J., pp. 46–48.

References

Ashby, E. and R. Wolf. 1947. A critical examination of the gravimtric method of determining suction force. *Ann Bot.* 11: 261–268.

Dainty, J. 1976. Water relations of plant cells. In Lüttge, U. and M. G. Pitman, (eds.), *Encyclopedia of Plant Physiology,* New Series, Vol. 2A. Berlin: Springer-Verlag.

Kramer, P. J. 1983. *Water Relations of Plants.* New York: Academic Press.

Meidner, H. and D. W. Sheriff. 1976. *Water and Plants.* New York: Wiley.

Measurement of transpiration rates with cobalt chloride paper

Introduction

The use of cobalt chloride paper represents one of the very early methods for assessing the rate of transpiration from the leaves of plants under conditions favoring transpiration. Though of historical value, it is a method very seldom used today.

With this method, transpiration is indicated by a change in color of dry paper disks impregnated with a slightly acidic solution of cobalt chloride. Dry paper disks are blue, but when exposed to humid air they gradually turn pink to red. Similarly, the impregnated paper will turn pink when exposed to a transpiring leaf surface. The rate of color change is indicative of the rate of transpiration. However, the true rate of transpiration is influenced by the disk when it is placed on the leaf surface by changes in air movement, a reduction in light reaching the covered stomates, and other effects.

Even though this method has serious limitations, it can be used with care to study the relative rate of transpiration under different conditions and to illustrate the kind of techniques scientists in the past devised to provide some understanding of physiological processes.

Purpose

To illustrate the preparation, standardization, and utilization of cobalt chloride paper in measuring water loss from the upper and lower surfaces of leaves from different plants; to determine the effect of light on the relative loss of water vapor from leaves.

Materials

10 plants (different species grown in the greenhouse)	5 filter paper strips (1 cm wide and 20 cm long)
4 bean plants (2 plants kept in light and 2 placed in darkness several hours before the experiment)	4 filter paper strips (1 cm wide and 5 cm long)
200 ml cobalt chloride solution (3%, w/v; 5.5 g $CoCl_2 \cdot 6H_2O$ + 94.5 ml of H_2O), slightly acidified with a few drops of HCl	squeegee board
	10 blotters
	oven (set at 40–50°C)
HCl (concentrated) in a dropping bottle	desiccator (with a dish of anhydrous calcium chloride in the bottom)
plastic tray (filled with water and covered with a wire screen)	forceps
	scissors
plastic tray (for soaking the filter paper strips)	10 petri dishes
	coverslips
	paper clips

Procedure

A. Preparation of cobalt chloride paper

Note: Prepare the paper before the laboratory meeting.

Prepare a 3% solution of cobalt chloride (5.5 g $CoCl_2 \cdot 6H_2O$ + 94.5 ml of H_2O) and slightly acidify with a few drops of HCl. Then soak sheets of filter paper (1 cm wide and 29 cm long) in the cobalt chloride solution until they are impregnated. Also treat several shorter strips of paper (1 cm wide and 4 cm long) in the same manner.

Roll the soaked strips gently on a squeegee board to remove excess solution. Partially dry the sheets between blotters in an oven at 40–50°C. When the edges of the paper begin to turn blue, remove from the oven and complete the drying process with an electric hand iron. Discard any papers that do not appear to be uniformly impregnated as judged by the uniformity of the blue color.

Place the acceptable strips into a desiccator containing calcium chloride and leave them for twenty-hour hours or until they are to be standardized.

B. Standardization of the cobalt chloride strips

Place a short strip of cobalt chloride paper (4 cm long and 1 cm wide) on a wire screen covering the top of a plastic tray containing water. The water level in the tray should be only a few millimeters from the lower surface of the wire screen. Allow sufficient time for the paper to change from blue to pink or white in color. Keep this strip on the wire screen and use it as a color standard to compare with the other strips.

Using forceps, quickly remove one cobalt chloride paper from the desiccator and place it on the wire screen beside the color standard. Note the time it takes for the strip to turn pink as compared to the color standard. Return the test paper to the desiccator. When it has retured to its original blue color, repeat the standardization process. Each cobalt chloride paper should be standardized several times. Also, remember to record the time necessary for the color change and calculate an average standarization time for each strip.

Color change (in seconds) and average standardization time:

After standardizing each cobalt chloride paper for the last time, cut it into twenty squares (each square 1 cm × 1 cm) and place the twenty squares into a petri dish labeled according to the average standardization time. Store the petri dishes containing the squares in a desiccator until ready to use.

C.   Relative transpiration rates from the upper and lower epidermis

Using forceps, remove several standardized cobalt chloride squares from the desiccator and place in contact with both surfaces of the leaves of several different species. Place a coverslip over each square and hold in place with a paper clip.

Record the time necessary for the papers to change from blue to pink in the same manner as for standardization. Using different squares of cobalt chloride paper, repeat the experiment several times. Record and average the time necessary for the color change of each paper. Whenever a complete color change does not take place within twenty minutes, record the transpiration rate as "negligible."

Observations:

D. Effects of light and darkness on loss of water vapor from plants

Place several potted and well-watered bean plants (or other species) in darkness overnight or for several hours before the laboratory meeting. Also maintain additional well-watered bean plants under light conditions.

Determine the relative rates of water vapor loss from the plants maintained in the light according to the cobalt chloride paper technique described in Part C. Make at least ten determinations from the upper and lower foliage surfaces and record the individual and average times for the cobalt chloride squares to change color.

After completing the determinations on the plants in light, remove the plants maintained in darkness to the laboratory and under subdued light conditions determine the relative rates of water vapor loss according to the same procedure as for the plants in the light. Record your results and average times.

Observations:

Results/
Conclusions

To arrive at a usable value for assessing the relative rates of transpiration for the plants studied, calculate the transpiration index (TI) as follows:

$$TI = \frac{\text{cobalt chloride paper standardization time (average time in seconds)}}{\text{cobalt chloride paper color change on leaf (average time in seconds)}}$$

The values obtained from the calculations for the different plants studied may be considered relative transpiration rates. The higher the resulting quotient from your calculations, the faster the rate of transpiration.

Present the results obtained in a suitable table indicating your observations and calculated values for each part of the experiment.

In concluding remarks, interpret your results with respect to the relative rates of water vapor loss from the upper and lower epidermis of a leaf, and the effect of light and darkness on the relative rate of transpiration. Consider the following questions in your conclusions:

1. Do some species of plants exhibit lower rates of transpiration from the upper than the lower epidermis? Explain.

2.   Can the absolute rates of transpiration be determined by the cobalt chloride paper method? Explain.

3.   What is the effect of light and darkness on the rate of water vapor loss from leaves?

4.   What are some of the pitfalls inherent in the cobalt chloride method for determining accurate transpiration rates from leaf surfaces?

Acknowledgment

The technique for preparation of the cobalt chloride paper was adapted from B. S. Meyer, D. B. Anderson, and C. A. Swanson. 1955. *Laboratory Plant Physiology*, 3rd edition. D. Van Nostrand Co., Inc., Princeton, N.J., pp. 164–165.

References

Bailey, L. F., J. S. Rothacher, and W. H. Cummings. 1952. A critical study of the cobalt chloride method of measuring transpiration. *Plant Physiol.* 27: 563.

Baker, D. A. 1984. Water relations. In Wilkins, M. B. (ed.), *Advanced Plant Physiology*. London: Pitman Publishing, Limited, pp. 297–318.

Dainty, J. 1963. Water relations of plant cells. *Adv. Bot. Res.* 1: 279–326.

Meidner, H. and D. W. Sheriff. 1976. *Water and Plants*. London: Blackie.

EXERCISE 13

Potometer method for studying transpiration

Introduction

The potometer method for measuring transpiration is actually based upon the rate of absorption of water. In many instances, however, the rate of absorption may be nearly equal to the rate of transpiration. To set up a potometer, a small shoot of a plant is sealed in a water-filled glass vessel. The glass vessel may have two outlets: a graduated capillary tube and a water reservoir. (In a homemade apparatus, the reservoir outlet is not necessary.)

The entire apparatus must be filled with air-free water and the system devoid of air spaces. An air bubble is then introduced into the capillary tube. When the plant transpires, the air bubble will move along the capillary tube and its rate of movement will give a measure of the rate of transpiration. Although the apparatus actually measures water absorption, under most conditions the rate of water absorption is almost equal to the rate of transpiration. However, if the transpiration rates become exceedingly high, it is possible that water absorption will lag behind water being lost by transpiration. Accordingly, it is possible that in some plants an absorption lag may be as much as 50%. For our purposes, however, the potometer can be used quite effectively for measuring the relative rates of transpiration under different conditions.

Purpose

To study the effects of various environmental factors upon transpiration as measured by the potometer method.

Materials

herbaceous plant (geranium, sunflower, bean, tomato, or *Fuchsia)*

wide-mouthed bottle and stopper

capillary tubing

1 set cork borers

1 liter water (boiled and then cooled to room temperature)

dishpan

bright light source

fan

Procedure

A.   The potometer apparatus

Prepare the potometer as illustrated in Figure 13–1 and according to the following directions:

1.   Select a suitable stopper that will fit a wide-mouthed bottle rather snugly. Drill a small hole in the stopper large enough to accommodate the end of a U-shaped capillary tube. Leave sufficient room to allow for another hole, the diameter of which is approximately that of the plant stem.

2.   Drill the second hole from the tapered end of the stopper with a cork borer that will just fit over the stem of a geranium or other suitable plant. Once the hole is drilled *do not* remove the cork borer.

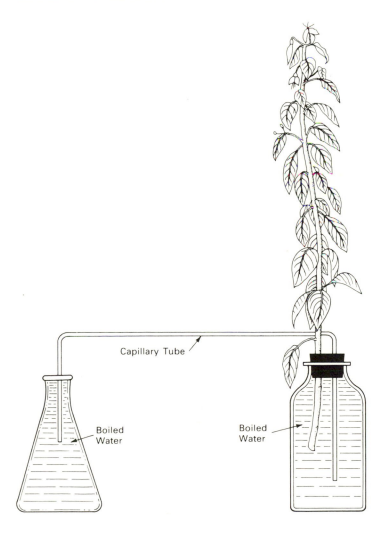

Capillary Tube

Boiled Water

Boiled Water

FIGURE 13–1   The potometer apparatus.

3. Cut a geranium plant under water (use a dishpan or sink) and insert the cut end of the leaf-bearing stem into the cork borer protruding from the stopper (also under water). The following operations should then be performed with the cut end of the stem, the stopper, and bottle under water.

4. Pull the stem through the stopper by withdrawing the cork borer. Quickly insert the stopper into the small wide-mouthed bottle filled with boiled water so that the stem extends one or two inches into the bottle. Do not push the stopper snugly into place.

5. Once the cut end of the stem is under water in the bottle, remove the entire apparatus from the sink and insert one end of the U-shaped capillary tube into the remaining hole of the stopper and about one inch into the bottle.

6. The capillary tube should also be filled with water. This may be accomplished by placing the open end of the tube into a beaker or flask containing previously boiled water and then pushing the stopper snugly into the bottle. This will force a sufficient amount of water from the filled bottle into the tubing and will also force the air out of the system. Keep the open end of the tubing in the beaker of water so that the system has no leaks. If necessary, seal any openings in the stopper with paraffin.

B. Operation of the potometer

When the system is completely filled with water, select a suitble distance along the horizontal part of the tubing and mark an origin and end point with a marking pencil.

Introduce a small bubble into the tubing by lifting the open end out of the beaker of water for a short time. Once a bubble of air is trapped, return the open end to the water. Use the bubble as an indicator of the rate of transpiration by timing how fast it moves over a set distance through the horizontal part of the capillary tube.

C. Effect of environmental factors on transpiration

Select various positions in the laboratory and the greenhouse, and by recording measurements of temperature, humidity, light intensity, wind, and rate of bubble movement, study the effect of environmental factors upon transpiration. At the end of the study on environmental factors, the plant could be sprayed with an antiperspirant spray and the effect noted.

Observations:

Results/
Conclusions

Present suitably illustrated results and justified conclusions as to the effect of various factors upon the rate of transpiration, as determined by the potometer method. Your conclusions should also be based on the answers to the following questions:

1.   Does the potometer method actually measure transpiration, and why is it not considered to be an accurate method? According to Maximov, the potometer method may be in error by as much as 50%. Explain.

2.   Why was air-free water used in the potometer?

3.   What conditions retard transpiration?

4.   Why is the rate of transpiration usually greater in sunlight than in the shade, even though the stomata may not be closed in the latter case?

References

Crafts, A. S., H. B. Currier, and C. R. Stocking. 1949. *Water in the Physiology of Plants.* Waltham, Mass.: Chronica Botanica Co.

Kramer, P. J. 1983. *Water Relations of Plants.* New York: Academic Press.

Martin, E. V. and F. E. Clements. 1935. Studies of the effect of artificial wind on growth and transpiration in *Helianthus annuus. Plant Physiol.* 10: 613.

Maximov, N. A. 1929. *The Plant in Relation to Water* (Trans. by R. H. Yapp). London: Allen and Unwin.

Slavik, B. 1974. *Methods of Studying Plant Water Relations.* New York.: Springer-Verlag.

EXERCISE 14

The path and rate of
water ascent in plants

Introduction

A major portion of the water taken up into the root hairs or other epidermal cells in or near the root hair zones moves from these cells into the cortex tissue, through the endodermis, the pericycle, into the xylem, and upward.

Much of the water taken up by the root hairs moves from the area of absorption along the walls of the cortical cells. Further movement along the endodermal cell walls is impeded, however, by the casparian strip, a band of suberin along the surface of the transverse and radial primary walls of the endodermal cells. Water is therefore diverted into the endodermal cells and flows along an osmotic gradient to the pericycle and into the conducting cells of the xylem. Since the xylem tissue of the roots directly connects with that of the stem, water moves out of the root and into the stem. The xylem of the stem consists of a complex network of water-conducting cells that lead to the fine veins or vascular bundles of the leaves. Water moves up this network into the veins, then from the tracheids of the xylem in the veins to the bundle cells (if present), and eventually into the mesophyll cells.

A small amount of the water may be used by the cells, although a large amount moves to and wets the mesophyll cell walls. Water then evaporates from a liquid film of water on the cell walls to vapor into the intercellular spaces. Water vapor is then lost from the spaces through open stomates or through the cuticle into the surrounding atmosphere.

Purpose

To study the rate at which an aqueous solution of acid fuchsin rises in plants, and to study the tissues involved in the translocation of the dye.

Materials

potted sunflower (or other herbaceous plant at least 50 cm tall)
100 ml acid fuchsin solution (0.25%, v/v)
small bottle

pan of water
razor blade
light source
microscope slides and coverslips
microscope

Procedure

Wash the soil from the roots of the potted sunflower plant. Immerse the entire root system in water and cut off the roots so that only the top 5 cm remains. After five minutes, transfer the plant and immerse the root in a 0.25% solution of acid fuchsin. Immediately cut off all except the top 2 cm of the roots. Maintain the plant in the dye in front of a bright light source.

Determine the time required for the dye to rise to a distance of 50 cm. The stem should be sufficiently translucent to permit direct observation of the rise of the dye.

Allow the dye to rise into the leaves and observe its movement into the vascular bundle of the leaf. Examine the blade of one of the leaves under a microscope and observe the completeness with which the dye has penetrated into the leaf vessels. Also, cut cross sections and longitudinal sections of the stems and roots and make a drawing indicating into what tissue the dye has penetrated.

Observations:

Results/ Conclusions

In your results, indicate the rate of movement of the acid fuchsin up the stem and present the drawings of the cross longitudinal sections of the stem studied.

Concluding remarks should include a consideration of the controlling influence on the rate of water movement in the stem. What does this exercise illustrate about the continuity of the water conduction system from the roots to the leaves? The rise of water in plants with intact root systems may not be as rapid as indicated in this exercise. Explain.

Acknowledgment

This exercise was adapted from B. S. Meyer, D. B. Anderson, and C. A. Swanson. 1955. *Laboratory Plant Physiology*, 3rd edition. D. Van Nostrand Co., Inc., Princeton, N.J., p. 52.

References

Anderson, W. P. 1976. Transport through roots. In Lüttge, U and M. G. Pitman (eds.), *Encyclopedia of Plant Physiology*. New Series, Vol. 2, Transport in Plants 2, Part B. New York: Springer-Verlag, p. 129.

Kramer, P. J. 1983. *Water Relations of Plants*. New York: Academic Press.

Milburn, J. A. 1979. *Water Flow in Plants*. New York: Longman, Limited.

Robards, A. W. and D. T. Clarkson. 1976. The role of plasmodesmata in the transport of water and nutrients across roots. In Gunning, B. E. S. and A.

W. Robards (eds.), *Intercellular Communications in Plants: Studies on Plasmodesmata*. Berlin, Heidelburg, New York: Springer-Verlag, p. 181.

Tyree, M. T. 1970. The symplastic concept: A general theory of symplastic transport according to the thermodynamics of irreversible processes. *J. Theor. Biol.* 26: 181–214.

Weatherby, P. E. 1963. The pathway of water movement across the root cortex and leaf mesophyll of transpiring plants. In Rutter, A. J. and F. H. Whitehead (eds.), *The Water Relations of Plants*. London: Blackwell Scientific, p. 86.

EXERCISE 15

The Münch pressure flow model

Introduction

The Münch pressure flow hypothesis is perhaps the most popular explanation for phloem transport of solutes. This hypothesis rests on the assumption that a pressure gradient exists between the cells supplying solute (source) and the receiving cells (sink). According to this theory, metabolites are carried passively in the positive direction of the gradient. The flow may be upward to developing reproductive structures, young expanding leaves, or meristematic areas. It may also be in a downward direction from the leaves (source) to the roots. It is important to note that the unidirectional flow of solutes in water through the sieve ducts is driven by a hydraulic (turgor) pressure gradient.

The physical system designed to illustrate the hypothesis consists of two differentially permeable sacs (dialysis tubing), one containing a solution of higher solute concentration than the other. Both sacs and the water containers in which they are immersed have open channels that offer little resistance to the flow of solutes and water. Since the system is closed and the walls of the sac are differentially permeable, more water will enter the sac with the high solute concentration with a concommitant increase in turgor pressure. The pressure will be transmitted throughout the system via the open channel between the two sacs. Thus, a circulating system is created. There is an induced flow from one sac (source) to the other (sink) with solute being carried along passively. Water moves out of the sink by virtue of the pressure developed and is recycled via the open channel between the water containers.

The evidence, in the form of the analogy between the physical system illustrated and pressure flow in the phloem of plants, is discussed in the textbooks cited in the references for this exercise.

Purpose

To study a physical system proposed by Münch to explain the unidirectional flow of solutes through the sieve tubes of plants.

Materials

200 ml sucrose solution containing iodine potassium iodide (dissolve 40 g sucrose, 10 g iodine, 20 g potassium iodide together in water and adjust the final volume to 200 ml)

200 ml starch solution (5%, w/v)

200 ml glucose solution (40%, w/v)

glucose "testape" (or Benedict's reagent)

4 strips dialysis tubing (7 in. long; soaked in water)

4 one-hole rubber stoppers

4 screw clamps

string

2 U-shaped glass tubes (internal diameter 3 to 4 mm; see Figure 15–1)

4 beakers (600 ml)

Procedure

A. The flow of iodine potassium iodide (I_2KI) reagent

Select a strip of dialysis tubing (about 7 in. long) that has been soaked in water. Twist one end, bend it back on itself, and secure tightly with string or a screw clamp. Open the other end by rubbing it between your fingers and fill the tubing to within one inch from the top with a 5% (w/v) soluble starch solution. Now insert the tapered end of a one-hole rubber stopper into the open end of the dialysis sac and tie the tubing tightly to the stopper so that there are no leaks. Insert one end of a U-shaped piece of glass tubing (about 3 to 4 mm internal diameter) through the hole in the stopper and into the starch solution. Make sure that the glass tubing fits rather snugly.

Select another piece of dialysis tubing and tie off one end as before. Fill the tubing to within an inch from the top with sucrose and I_2KI solution. In the same manner as before, stopper the open end of the dialysis tubing and insert the free end of the U-shaped tube through the stopper and into the sucrose–I_2KI solution. Make sure there are no leaks in either of the sacs at each end of the U-shaped tube.

As indicated in Figure 15–1, support the entire system with a ring stand and clamp so that each sac is immersed completely in a separate beaker of water. Make observations during and at the end of the next two hours.

Observations:

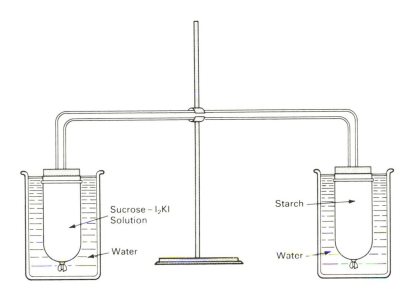

FIGURE 15–1   Apparatus for demonstrating the Münch pressure flow.

B.   The flow of glucose

Prepare the same model as outlined for Part A, but this time fill one dialysis tubing sac with glucose solution and the other with distilled water. Support this system as before with each dialysis sac immersed in a separate beaker of water. Maintain the system for twenty-four hours, then remove the sacs and test the contents of each for glucose. To test for glucose, use glucose "testape" by dipping a strip of testape into the solution. Remove the testape and wait for a color change. Compare the color of the tape with the color chart provided.

If glucose testape is not available, a standard Benedict's test may be performed according to the procedure outline in Exercise 19 (Part E). Your instructor may also wish to demonstrate the presence of glucose by means of gas chromatographic analysis.

Observations and glucose determinations:

Results/
Conclusions

Report your results and observations. In concluding statements, consider the following questions:

1.   What is the influence of turgor pressure on the flow of potassium iodide–iodine reagent and glucose from one sac to the other?

2.   How is the system studied in this exercise analogous to the translocation system of a living plant?

3.   Why is the model studied in this exercise useful in explaining the translocation of osmotically active organic substances within plants?

4.   What are some of the objections to the idea that the mechanisms of organic solute translocation are totally analogous to the Münch pressure flow model?

5.   Is the design of the mass pressure flow model used in this exercise different from that originally designed by Münch? Explain.

References

Canny, M. J. 1983. Translocation of nutrients and hormones. In Wilkins, M. B. (ed.), *Advanced Plant Physiology*. London: Pitman Publishing, Limited.

Curtis, O. F. 1936. *The Translocation of Solutes in Plants*. New York: McGraw-Hill Book Co.

Gunning, B. E. and A. W. Roberts (eds.). 1976. *Intercellular Communications in Plants: Studies in Plasmodesmata*. New York: Springer-Verlag. Moorby, J. 1981. *Transport Systems in Plants*. New York: Longman Group, Limited.

Münch, E. 1931. *Die Stoffbewegungen in der Pflanze*. Stuttgart: Gustav-Fisher-Verlag.

Peel, A. J. 1974. *Transport of Nutrients in Plants*. New York: John Wiley and Sons.

Scholander, P. F., H. T. Hammel, D. Bradstreet, and E. A. Hemmingsen. 1965. Sap pressure in vascular plants. *Science* 148: 339–346.

Zimmerman, M. H. 1960. Transport in the phloem. *Ann. Rev. Plant Physiol.* 11: 167–190.

EXERCISE 16

Stomatal aperture measurement with a diffusive resistance meter

Introduction

Perhaps one of the most reliable and easily operated instruments for measuring stomatal aperture is the diffusive resistance meter assembly. This equipment consists of (1) a chamber containing a sensor element, (2) a drying tube assembly for drying the sensor element, (3) a thermistor for measuring leaf temperature, (4) a meter that receives output from the sensing element, and (5) a stopwatch.

The sensor, constructed of lithium chloride and housed in the chamber that attaches to a leaf, exhibits progressive decreasing electrical resistance as the relative humidity inside the chamber increases. If the stomates are open, the relative humidity increases rapidly with a concommitant rapid decrease in electrical resistance. This decrease can be observed as the meter needle moves quickly from low to higher current flow. Slow needle movement indicates that the stomates are closed, thus providing a high diffusive resistance to water vapor. To determine rather precise measurements of the amount of water lost from a given leaf area and hence stomatal aperture, a stop watch is used to time the movement of the meter needle from low to a high current. If the rate of movement is slow, the diffusive resistance is high, transpiration is low, and the stomata are therefore closed.

Purpose

To gain insight in measuring stomatal aperture with the diffusive resistance meter.

Materials

diffusive resistance meter (Lambda
   Instruments Model LI60)
meter drying assembly with freshly
   dried desiccant
sensor cap assembly

ring stand, clamps, and rings
plants: *Coleus*, jimsonweed,
   tomato, and sunflower
stopwatch

Procedure

Note: Before operating the diffusive resistance equipment, be sure that your instructor has explained fully the procedure to be followed.

Connect the meter drying assembly and the sensor according to the directions given by your instructor. Turn the selector switch to *CAL* position, then adjust until a reading of 100 is obtained. If you are unable to reach 100 on the scale, replace the batteries. Then turn switch to position *Hum L* and pump dry air through the sensing device. Continue this until a meter reading of 14 or less is obtained. Holding the sensing device by the handle, attach it to the leaf to be studied, supporting the sensor so the leaf and petiole are not strained. Be certain that heat from the supporting hand does not warm the leaf or sensor. Do not let water touch the sensing device.

As the needle passes 20 on the meter, use the stopwatch to measure the time it takes the needle to reach and pass 60; the time in seconds should be noted. Record the temperature (in microamps) by switching to *Temp*. To prepare for studying the next plant type, remove the sensing device and pump dry air through it until the meter needle falls below 14.

Complete the following table of values using the plants provided by the instructor:

Time	Species	Upper Surface	Lower Surface
1			
2			
3			
4			

To convert the elapsed time values (Δt) to leaf resistance values, use the graphs provided by the instructor. Use the following steps:

1. Convert the temperature reading from microamps to degrees C by using the temperature calibration curve. Use the right side of the calibration curve provided.

2. From the left-hand scale of the same calibration curve, record the conversion factor.

3. Multiply the Δt by the conversion factor to calculate the Δtc.

4. Using the graph provided by the instructor, find the leaf resistance corresponding to the Δtc.

Results/ Conclusions

Based upon the idea that the lower the diffusive resistance the greater the amount of water vapor lost, which in turn reduces the resistance in the sensor, explain and discuss the differences in diffusive resistance between the upper and lower epidermis and between the species selected by the instructor. Why could direct breathing on the leaves affect the results of the experiment? What would be the probable effect of light or darkness on stomatal aperture?

References

Boyer, J. S. 1969. Measurement of the water status of plants. *Ann. Rev. Plant Physiol.* 20: 351–364.

Holmgren, P., P. G. Jarvis, and M. S. Jarvis. 1965. Resistance to carbon dioxide and water vapor transfer in leaves of different plant species. *Physiol. Plantarum* 18: 557–573.

Husken, D., E. Stendle, and U. Zimmermann. 1978. Pressure probe technique for measuring water relations of cells in higher plants. *Plant Physiol.* 61: 158–163.

Monteith, J. L. and T. A. Bull. 1970. A diffusive resistance porometer for field use. II. *J. Appl. Ecol.* 7: 623–638.

Slavik, B. 1974. *Methods of Studying Plant Water Relations.* New York: Springer-Verlag.

Van Bavel, C. H., F. S. Nakayama, and W. L. Ehaler. 1965. Measuring transpiration resistance of leaves. *Plant Physiol.* 40: 535–540.

EXERCISE 17

Plant nutrition and mineral deficiencies

This exercise is designed to illustrate the requirements of plants for several elements (termed *essential elements*) that are necessary for plant growth and development, reproduction, and survival.

The mineral elements essential for the successful development of a majority of plants are carbon, hydrogen, oxygen, nitrogen, phosphorus, potassium, calcium, sulfur, magnesium, and iron. The following elements—manganese, zinc, boron, copper, and molybdenum—are also known to be required by a majority of plants, but in trace amounts. Additional trace elements required for normal development in only some plants are sodium, aluminum silicon, chlorine, gallium, and cobalt. With the exception of carbon and oxygen, which are absorbed by plants in the form of carbon dioxide and oxygen gas, and other elements are normally absorbed from the soil.

In order to observe the effects of various elements on plant growth, plants are often grown in artificial media or washed white sand to which a solution of all the essential elements are added except for the element being tested. Another method is to grow plants in liquid culture (termed *hydroponics*) containing a balance of elements less the element(s) being tested. Hydroponics is used today in the commercial production of various crops.

It is important to note that experiments based upon liquid culture may show extreme deficiency systems when one element is lacking and the others are at optimum levels. At optimum levels the available elements tend to enhance the biochemical reactions for which they are essential and inhibit reactions that require the deficient element(s). Conversely, under field conditions, a deficiency of one element is not always readily apparent because several elements may be low. Nevertheless, symptoms of elemental deficiencies often help farmers make proper application of fertilizer, especially when observations in the field are correlated with the levels of elements in plants as determined by laboratory analysis. In addition, plant scientists often use hydroponic systems and artificial media to study the role of elements in plant metabolism and growth and the pattern of translocation of certain elements. For example, some elements are considered mobile and will be translocated to younger leaves when that particular element becomes lacking, while other elements are immobile and remain in the older leaves. In the former case, the deficiency symptoms are manifested in the older leaves, whereas the opposite is true with immobile elements (that is, symptoms observed predominantly in the younger leaves).

Purpose

To demonstrate the mineral deficiency symptoms exhibited by plants grown in various nutrient solutions.

Materials

40 bean seedlings (*Phaseolus vulgaris* L.), 7 to 10 days old

13 aqueous stock solutions in glass bottles (Table 17–1)

12 volumetric flasks (1 liter) for preparation of stock solutions

13 pipets (5 ml)

several large flasks and beakers

11 one-quart jars (or ceramic crocks)

11 plastic bags

11 fiberboard covers with four holes

30 corks (with a hole in each initially and then split lengthwise into two equal halves)

1 set cork borers

razor blade

scalpel

marking pencil

cotton

greenhouse space

ruler

Procedure

A. Plant material

Various species of plants may be used to demonstrate mineral deficiencies. However, tomato, sunflower, tobacco, or bean plants are usually used since they manifest the various symptoms very well. To obtain seedlings of the desired stage (seven- to ten-day-old seedlings), sterilize seeds by immersing them twenty minutes in commercial hypochlorite solution which had been diluted 1:6 with distilled water. If the seeds tend to float, they may be covered with absorbent cotton saturated with the solution.

After sterilization, place the seeds on moist vermiculite in pots or flats and cover with one-half inch of additional vermiculite. The seeds should germinate readily at room temperature or in the greenhouse. To maintain the germinating seedlings, use *only distilled water*.

B. Stock solutions

Prepare the stock solutions by putting the amount of each chemical (Table 17–1) into a flask and adding distilled water to volume (1 liter). Use only one chemical per container. If less than one liter of stock solution is required, reduce the amount of the chemical proportionately.

TABLE 17–1   Stock solutions

Chemical	Formula	Amount (Grams/Liter)	Molarity (M)
A. Ammonium acid phosphate	$NH_4H_2PO_4$	23	0.20
B. Ammonium nitrate	NH_4NO_2	40	0.50
C. Calcium nitrate	$Ca(No_3)_2$	189	1.15
D. Calcium chloride	$CaCl_2$	29	0.26
E. Magnesium chloride	$MgCl_2 \cdot 6H_2O$	41	0.20
F. Magnesium nitrate	$Mg(NO_2)_2 \cdot 6H_2O$	51	0.20
G. Magnesium sulfate	$MgSO_4 \cdot 7H_2O$	99	0.40
H. Potassium acid phosphate	KH_2PO_4	27	0.20
I. Potassium nitrate	KNO_2	121	1.20
J. Potassium sulfate	K_2SO_4	87	0.50
K. Ferric chloride	$FeCl_2 \cdot 6H_2O$	10	0.04

			Molarity ($\times 10^{-2}$)
L. Microelement stock solution elements mixed/1 liter distilled water:			
Boric acid	H_2BO_2	0.72	1.200
Copper chloride	$CuCl_2 \cdot 2H_2O$	0.02	0.012
Manganese chloride	$MnCl_2 \cdot 4H_2O$	0.45	0.230
Zinc chloride	$ZnCl_2$	0.06	0.044
Molybdic acid (85% MoO_4)	$H_2MoO_4 \cdot H_2O$	0.01	0.006

M. *Fe EDTA* (iron complex of ethylenediaminetetraacetic acid):
Dissolve 1340 mg disodium ethlenediaminetetraacetate
($Na_2C_{10}H_{14}O_3N_2 \cdot 2H_2O$) in 500 ml of distilled water and heat. While still hot add 990 mg $FeSO_4 \cdot 7H_2O$ and stir vigorously.

C.   Containers and covers

Various types of containers, such as ice cream cartons, crocks, or mason jars, covered with light-excluding aluminum foil, brown paper, or other materials, may be used. Regardless of the container used, they must be washed thoroughly before use or lined with plastic bags to hold the nutrient solutions and rather effectively ensure excellent results.

Square covers (of fiberboard or cardboard impregnated with paraffin) with four equally spaced holes may be provided. If covers are not provided, they may be made by impregnating a sheet of brown wrapping paper with hot paraffin and sticking it to a similarly impregnated sheet of white bond paper. Then punch out four equally spaced holes in each cover.

D.   Preparation of nutrient solutions

Wash eleven one-quart jars with detergent and rinse several times with tap water. Wrap the outside of each jar with aluminum foil so that the shoulder and neck are completely covered. Next, line each jar with a plastic bag (large size bags are available if crocks are to be used).

Prepare the various nutrient solutions (specific nutrient solutions for each student or group will be designated by the instructor) by filling the one-quart jars one-half full of *distilled* water followed by the addition of stock solution (in milliliters) as indicated in Table 17–2. The number of milliliters of each stock solution given is the amount required for a liter of nutrient solution. If a container larger than one quart is used, make the solution volume adjustment accordingly. Each stock solution should be added slowly and mixed thoroughly before the next is added. *Do not* contaminate the stock solutions or else poor results can be expected.

TABLE 17–2   Nutrient solutions

Stock Solutions	None	Plants with Deficiency Symptoms of *								
		N	P	K	Ca	Mg	S	Fe_1	Fe_2	Micro-elements
A	5	—	—	5	5	—	—	5	5	5
B	—	—	1	6	8	6	—	—	—	—
C	5	—	5	5	—	5	5	5	5	5
D	5	21	5	5	—	5	—	5	5	5
E	—	—	—	—	—	—	5	—	—	—
F	—	—	—	—	—	—	5	—	—	—
G	5	5	5	5	5	—	—	5	5	5
H	—	5	—	—	—	5	5	—	—	—
I	5	—	5	—	5	1	5	5	5	5
J	—	5	—	—	—	4	—	—	—	—
K	—	—	—	—	—	—	—	—	2	—
L	2	2	2	2	2	2	2	2	2	—
M	2	2	2	2	2	2	2	—	—	2

* The numbers indicated represent the amount in milliliters of the stock solution to be used per liter of test solution.

After the stock solutions are added, fill the jars to the top with *distilled water*. Fill the remaining empty jar with *distilled water only*. This will serve as a control (final volume is approximately one liter). Cover and label each jar as soon as it is filled.

E.   Transfer of seedlings to nutrient solutions

Just before transferring the seedlings, the vermiculite should be thoroughly soaked with distilled water so that the plants may be removed with the least possible injury to the roots. The roots should be rinsed clean with distilled water and the plants immediately placed in the nutrient solution.

Place three seven- to ten-day-old bean seedlings in each jar, one through each hole in the cover, retaining the fourth hole for watering. Prop the seedlings erect with loose wads of cotton wrapped around the stems and half pieces of cork so that the root system is immersed in the liquid. Immediately fill the jars to just below the lid with distilled water. Stir each solution with a clean glass rod and place the jars in the allotted greenhouse space.

For the first few days, damaged or infected seedlings should be replaced with fresh ones from the germination trays. Later casualties should not be replaced, but should be accounted for in reporting the experiments.

Examine the cultures at least three times a week, each time adding distilled water as needed. At the same time, bubble air through the solution. When the roots are well developed, the solution level may be allowed to drop a little.

F. Seedling growth and observations

Make weekly observations and take notes on the appearance of the plants. Record the appearance of the plants after three and five weeks' growth (Table 17–3). Since the symptoms connected with a specific deficiency are usually different among species, there is no precise description of symptoms covering all plants.

TABLE 17–3   Observed mineral deficiency symptoms

Deficient Element	Observations—3 weeks	Observations—5 weeks
None		
Nitrogen (N)		
Phosphorus (P)		
Potassium (K)		
Calcium (Ca)		
Magnesium (Mg)		
Sulfur (S)		
Iron (Fe_1)		
Iron (Fe_2)		

Nevertheless, the most commonly observed symptoms exhibited by bean plants growing in nutrient medium lacking one of the following elements are presented: phosphorus, nitrogen, calcium, magnesium, potassium, iron, sulfur (Table 17–4). On the basis of your observations, write in the appropriate deficient element that best corresponds to the given symptom.

At the conclusion of the experiment (about five weeks after planting), harvest the plants from each jar. Record length of tops, length of roots, wet weights of tops, wet weights of roots, dry weight of tops, and dry weight of roots. If desired, set aside several intact plants for ashing and elemental content determinations (see Exercise 18).

TABLE 17–4   General characteristics of deficiency symptoms

Symptoms	Deficient Element
1. Plants generally pale green, although some plants may appear reddish in color; lower leaves generally yellow, dry to light brown, and defoliate; stunted growth with stalks short and slender.	_____
2. Plants usually dark green, but often developing red and purple color; lower leaves sometimes yellow, drying to greenish brown or black color; early defoliation; dwarfed plants with thin shoots and small leaves.	_____
3. Lower leaves mottled or chlorotic (interveinal chlorosis of oldest leaves); leaves may have dead spots, may redden, and have tips and margin turned or cupped upward.	_____
4. Oldest leaves (lower) mottled or chlorotic with small spots of dead tissue, usually at tips and between veins; marginal browning of leaves; stalks slender; forward curling of leaves.	_____
5. Yellow terminals with oldest leaves remaining green; stunted growth, die back of terminals, abnormal growth of young shoots, followed by die back, pitted stems, rot of stem centers; short stubby roots sometimes with black spots; roots slimy in appearance and to touch.	_____
6. Young leaves chlorotic, net-like interveinal chlorosis, followed by browning of leaves on young growth; principal veins typically green; stalks short and slender.	_____
7. Terminal bud commonly remains alive; wilting or chlorosis of young or bud leaves; young leaves with veins and tissue between veins light green in color.	_____

Measurement of plants:

Results/
Conclusions

Present your results based on the observations recorded in Table 17–3 and the measurements of the plants taken at the end of the exercise.

As a guideline to an interpretation of the results, list the known physiological roles of the various elements listed in this exercise. Draw general conclusions as to the means by which an investigator may be able to distinguish symptoms arising from mineral deficiencies as opposed to those arising from parasite infection. What are some of the advantages and disadvantages of using hydroponics for the study of mineral nutrition of plants?

Acknowledgment

This exercise was adapted from the work of J. O. Dutt and E. L. Bergman.

References

Alexander, L. J., V. H. Morris, and H. C. Young. 1939. Growing plants in nutrient solution. *Ohio Agric. Exp. Sta. Spec. Circ.* 56.

Clarkson, D. T. and J. B. Hanson. 1980. The mineral nutrition of higher plants. *Ann. Rev. Plant Physiol.* 31: 239–298.

Crafts, A. S. and T. C. Broyer. 1938. Migration of salts and water into xylem of the roots of higher plants. *Amer. J. Bot.* 25: 529–535.

Dutt, J. O. and E. L. Bergman. 1966. Nutrient solution culture of plants. *Penn. State Agric. Ext. Ser. Veg. Crops* 2.

Epstein, E. 1972. *Mineral Nutrition of Plants: Principles and Perspectives.* New York: Wiley.

Hoagland, D. R. and D. I. Arnon. 1950. The water-culture method for growing plants without soil. *Calif. Agric. Exp. Sta. Circ.* 347.

Smith, J. W. and W. R. Robbins. 1938. Methods of growing plants in solutions and sand cultures. *N. J. Agric. Exp. Sta. Bull.* 636.

Stout, T. G. and M. E. Marvel. 1959. Hydroponic culture of vegetable crops. *Florida Agric. Ext. Ser. Circ.* 92.

Wallace, T. 1961. *The Diagnosis of Mineral Deficiencies in Plants*, 3rd edition. New York: Chemical Publishing Co., Inc.

Macroelements
in plant ash

Introduction

The analysis of mineral elements in plants has been important in understanding the localization and role of elemental nutrients in plants. Although modern methods of plant analysis depend upon ashing plant structures, the extracts are often generally analyzed with an emission spectrometer. In some cases, however, wet chemistry tests are still relied upon heavily.

Analyses of plants with respect to elemental content enable extension specialists to make reasonably accurate recommendations to home and commercial growers for the proper application of fertilizers and adjustments of soil pH. For these reasons, the methods of dry ashing and detection of elements in plant ash are provided. Since most of the individual test procedures are self-explanatory, further classification is not necessary.

Purpose

To determine qualitatively some of the macroelements in plant ash by standard chemical techniques.

Materials

50 g fresh plant material (seeds, tuber tissue, stems, or leaves)

100 ml of the following separate 5% (w/v) solutions in drop bottles: KH_2PO_4, K_2HPO_4, $MgSO_4$, K_2SO_4, $CaCl_2$, KSCN

100 ml of the following separate solutions in bottles: 15% (v/v) $HClO_4$, 10% (w/v) $BaCl_2$, 50% (v/v) H_2SO_4

ammonium molybdate reagent (see Appendix II)

magnesium reagent (see Appendix II)

diphenylamine–concentrated sulfuric acid reagent (see Appendix II)

large crucible (or aluminum foil cup)

drying oven (set at 100 to 105°C)

analytical balance

muffle furnace (set at 550°C)

watchglass or spot plate

microscope

microscope slides and coverslips

Procedure

A. Method of dry ashing

Place about 50 g of fresh plant material (seeds, tuber tissue, stems, or leaves) into a preweighed container (large crucible or aluminum foil cup). Dry the material in an oven set at 100 to 105°C for twenty-four hours or until there is no change in weight. Then weigh the container plus the sample again and calculate the dry weight of sample.

Grind the dried material (clean mortar and pestle or mill) and transfer it to a tared crucible or porcelain dish. After recording the weight, put the dish into a muffle furnace and ash at 550°C for two hours or until a constant weight is obtained. After ashing, cool and reweigh the crucible and contents and record all of your measurements as follows:

Plant part and species used:

Fresh and dry weight determinations   Weight in grams:
    Container plus sample
    Container
    Sample fresh weight
    Container plus dried sample
    Container
    Sample dry weight
    Percent dry weight of fresh sample

Ash determination   Weight in grams:
    Crucible plus dried sample
    Crucible
    Crucible plus ash
    Crucible
    Sample ash weight
    Percent ash of dry weight
    Percent ash of fresh weight

B. Detection of macroelements in plant ash

Dissolve the ash in 4 ml of distilled water (the addition of a *very* small amount of HCl may be required). Use this solution in the following tests:

Phosphorus

Place a drop of 5% KH_2PO_4 solution on a clean microscope slide and next to it a drop of ammonium molybdate reagent. Draw one of the drops into the other with the edge of a coverslip and wait several minutes. Using a microscope, observe and note the color and form of the crystals found. Repeat the same procedure for one drop of ash solution.

Observation:

Potassium

Place a drop of a 5% of K_2HPO_4 solution on a slide and next to it a drop of 15% $HClO_4$ (*Caution!*). Following the same method as for phosphorus, mix the drops and look for the formation of $KClO_4$ crystals. After noting and recording their characteristic shape and color, perform the same test on a drop of ash solution.

Observation:

Magnesium

Test a drop of 5% $MgSO_4$ with a drop of magnesium reagent in the same manner as above and observe the ammonium–magnesium–phosphate crystals. Test the ash solution.

Observation:

Sulfur

Place a drop of 5% K_2SO_4 and a drop of 10% $BaCl_2$ on a glass microscope slide. After mixing the drops, note the crystals of $BaSO_4$ that indicate the presence of sulfur. Repeat the test for the ash solution.

Observation:

Calcium

Test a drop of 5% $CaCl_2$ with a drop of 50% H_2SO_4. Observe the formation of $CaSO_4$ crystals. Test the ash solution in the same way.

Observation:

Nitrogen

To a clean watchglass or spot plate, add five drops of ash solution. Add a drop of 1% diphenylamine–H_2SO_4 reagent and note the development of color, which indicates the presence of nitrates.

Observation:

Iron

Add five drops of ash solution and three drops of 5% KSCN to a clean spot plate. The formation of a red color indicates the presence of iron.

Observation:

Results/ Conclusions

From the analysis of ash, construct a table listing the elements tested, the plant material(s), and the test used. Appropriately indicate a positive or negative test with the color and/or precipitate or crystals formed.

Your conclusions should be based on the following considerations and questions:

1. During the process of ashing, what elements are lost by volatilization?

2. Would the dry ashing techniques used in this exercise be suitable for the quantitative determination of phosphorus and sulfur? Explain.

3. Is the presence of large amounts of a particular element indicative of the plant's nutritional requirements?

4. What is the significance of high levels of elements in seeds?

5. What are the limitations of the techniques used in this exercise, and of what pitfalls must investigators be aware?

References

Carolus, R. L. 1938. The use of rapid chemical plant nutrient tests in fertilizer deficiency diagnosis and vegetable crop research. *Va Truck Exp. Sta. Bull.* 98

Chapman, H. D. and P. F. Pratt. 1961. Methods of analysis for soils, plants and waters. Berkeley: University of California Division of Agricultural Science.

Cheng, K. L. and R. H. Bray. 1951. Determination of calcium and magnesium in soil and plant material. *Soil Sci.* 72: 449–458.

Fiske, C. H. and Y. Subbarow. 1925. The colorimetric determination of phosphorus. *Biol. Chem.* 66: 375–400.

Johnson, C. M. and A. Ulrich. 1959. Analytical methods for use in plant analysis. *Calif. Agri. Exp. Sta. Bull.* 766

Piper, C. S. 1942. *Soil and Plant Analysis: A Laboratory Manual of Methods for the Examination of Soils and Determination of the Inorganic Constituents of Plants.* Adelaide: University of Adelaide.

Piper, C. S. 1944. *Soil and Plant Analysis*. New York: Interscience Publishers, Inc.

Thornton, S. F., S. D. Conner, and R. R. Fraser. 1939. The use of rapid chemical tests on soils and plants as aids in determining fertilizer needs. *Rev. Ed. Ind. Agri. Exp. Sta. Circ.* 204.

EXERCISE 19

Simple tests for carbohydrates

Carbohydrates are important to plants in several major ways. First, they provide the carbon skeleton for the functional and structural compounds in plants. Second, they are important constituents of the structural tissues. Third, they provide the energy used to drive the metabolic reactions.

The three major categories of carbohydrates are (1) monosaccharides, (2) oligosaccharides, and (3) polysaccharides. The monosaccharides are considered to be the simplest form of carbohydrate and are classified according to the number of carbon atoms present. Hence, monosaccharides include the trioses (three carbon atoms), pentoses (five carbon atoms), hexoses (six carbon atoms), and so on. Monosaccharides may also be classified on the basis of whether one of the carbons bears a carbonyl oxygen as part of an aldehyde group (aldoses) or as a ketone (ketoses). These groups are called *reducing groups* because they are readily oxidized by compounds that are readily reduced. Sugars possessing these groups are therefore called *reducing sugars*.

Oligosaccharides are generally classified according to the number of monosaccharide units found in their structure. For example, if two monosaccharides make up a sugar, it is termed a disaccharide; if three, a trisaccharide; if four, a tetrasaccharide, and so on. One of the major disaccharides found in higher plants is sucrose, a condensation product of glucose and fructose. Although free glucose or fructose are reducing sugars, sucrose itself is not because the reducing groups of both simple sugars are involved in the bond that links them together.

Other oligosaccharides of importance in plants are maltose (disaccharide), a degradation product of starch, cellobiose (disaccharide), a degradation product of cellulose or lignin, gentianose (trisaccharide), and raffinose (trisaccharide). The tetrasaccharide stachyose found in several tree species, upon hydrolysis yields glucose, fructose, and two molecules of galactose.

Unlike the oligosaccharides, polysaccharides are high molecular weight polymers of monosaccharide-repeating units. The two most common polysaccharides of plants are (1) starch, a storage product of plants, and (2) cellulose, the structural polysaccharide that makes up the greater part of the plant cell wall. Starch consists of α-1,4-, α-1,6- and sometimes α-1,3-linkages of glucose. Conversely, cellulose consists of β-1,4-linkages of glucose. For further details concerning the chemistry of carbohydrates, consult the references listed at the end of this exercise.

The current procedures for determining the kinds and amounts of mono- and oligosaccharides in plants involve the use of gas chromatography and high-performance liquid chromatography equipment. These instruments provide for the separation, identification, and quantitation of sugars extracted from plants. If such equipment is available, your instructor may demonstrate its operation. The following exercises, however, are provided to demonstrate some of the chemical procedures that have been used in the past and have provided plant scientists with much of our present information concerning the chemistry of carbohydrates, their relative levels in plants, and their possible biochemical and physiological roles. A brief discussion of many of these tests follows.

The Molisch test is a general test for simple carbohydrates, (that is, non-polymeric forms). The acid used (H_2SO_4) has two purposes. First, it may hydrolyze the oligosaccharide that might be present in the test solution and reacts with monosaccharides to produce furfuryl derivatives, which in turn react with the α-naphthol to give a colored product.

Barfoed's test is used to distinguish between monosaccharides and disaccharides under controlled conditions of pH and temperature. The test is based on the fact that monosaccharides reduce cupric ions more readily in a weak acid solution than do the disaccharides. Therefore, the monosaccharides will indicate their presence with the formation of a red cuprous oxide precipitate when treated with Barfoed's reagent.

Bial's test is specific for pentoses, which are dehydrated in the presence of HCl to a furfuryl derivative. Furfuryl and orcinol then condense in the presence of ferric chloride to produce a colored product.

Seliwanoff's test is a timed color test for ketoses in which they become dehydrated faster than aldoses to form furfuryl. The furfuryl then complexes with resourcinol to yield a green to deep blue color. This is a timed test (time of heating), since other sugars will give positive tests if the time of heating is relatively longer than twenty minutes.

Benedict's test is specific for reducing sugars and is based on the reducing action of ketones and aldehydes with oxidation to their corresponding acids (for example, glucose to glucuronic acid). Further, cupric ions present in the reagent are reduced with the formation of the brick red precipitate, cuprous oxide, as follows:

$$\begin{array}{c} CHO \\ | \\ R \end{array} + 2Cu^{++} + 5OH \rightarrow R + COO + Cu_2O + 3H_2O$$

The basic conditions of the reagent stabilize the cupric ion and prevent it from precipitating as black insoluble cupric oxide.

The phenylhydrazine test is a very old one; it is based on the reaction of some sugars with the reagent to form crystalline osozones. The crystals formed by reactive sugars can be observed with a microscope. Further, they have characteristic shapes for some of the reacting sugars. The crystals of some sugars may be quite similar, while other sugars react to form very distinct crystal shapes.

The final exercise in this series demonstrates the enolization of sugars. Enolization is the migration of a proton from one carbon atom to the oxygen atom of an adjacent carbonyl group with the formation of an unsaturated alcohol (enol form). The reaction may be illustrated accordingly:

$$\underset{\text{}}{\overset{\text{O}\quad\text{H}}{\text{C}-\text{C}}}\ \overset{\text{H}}{\underset{\text{}}{}}\ \xrightleftharpoons{\text{OH}}\ \underset{\text{enol}}{\overset{\text{OH}\quad\text{H}}{\text{C}=\text{C}}}$$

or

$$\underset{\text{}}{\overset{\text{O}\quad\text{OH}}{\text{C}-\text{C}}}\ \overset{\text{H}}{\underset{\text{}}{}}\ \xrightarrow{\text{OH}}\ \underset{\text{enediol}}{\overset{\text{OH}\quad\text{OH}}{\text{C}=\text{C}}}$$

It is through the enolization reaction that aldoses can ultimately be converted to ketoses, accounting for the interconversions of some of the monosaccharides.

Purpose

To demonstrate the use of classical tests for the detection of simple carbohydrates present in plant tissues.

Materials

different plant materials: apples, celery petioles, carrots, grapes, onions, turnips, etc.

7 separate 1% (w/v) sugar solutions of glucose, arabinose, fructose, maltose, lactose, sucrose, and ribose

6 separate test reagents (see Appendix II): Molisch, Barfoed's, Bial's, Seliwanoff's, Benedict's, and phenylhydrazine

30 ml H_2SO_4 (concentrated)

60 ml HCl (concentrated)

10 ml HCl (0.1N)

5 ml preservative (1%, v/v thymol in toluene)

10 ml $Ba(OH)_2$ saturated solution

7 test tubes for each test

microscope

slides and coverslips

litmus paper

mortar and pestle

Procedure

A. Molisch test

Label eight test tubes with the initial letter of glucose, arabinose, fructose, maltose, lactose, sucrose, ribose, and water. To each tube add 5 ml of the corresponding one percent sugar solution or water (control). Mix two drops of Molisch reagent with each sugar solution and control. Then incline each tube and carefully pour inside 2 ml of concentrated H_2SO_4. The red-violet color formed at the contact zone indicates the presence of a carbohydrate. Compare the relative extent of the reaction in the tubes and record your results in Table 19–1. Repeat the test on the plant material provided (see Part H) and record the results.

B. Barfoed's test

Boil in test tubes 5 ml of Barfoed's reagent plus 1 ml of each one percent sugar solution. Use 1 ml of water plus 5 ml of reagent as a control. Allow to stand for four to five minutes and compare the tubes. A positive test is indicated by a red cuprous oxide precipitate. Record your results in Table 19–1. Repeat the test on the plant material provided and record the results.

C. Bial's test

Pipet 4 ml of each 1% sugar solution into separate test tubes. In addition, set up a tube with 4 ml of distilled water. Add 2 ml of Bial's reagent followed by 5 ml of concentrated HCl to all tubes. Mix and stopper the tubes with cotton and heat in a boiling water bath for ten minutes. A positive reaction is indicated by the development of a green to deep blue color. Record the results in Table 19–1. Repeat the tests on the plant material provided and record the results.

D. Seliwanoff's test (timed color test specific for ketoses)

Place 3 ml of Seliwanoff's reagent in each of eight test tubes labeled with the initial letter of glucose, arabinose, fructose, maltose, lactose, sucrose, ribose, and water. To each tube add three drops of a one percent solution of the corresponding sugar or water.

Place the eight tubes simultaneously in boiling water and heat until a deep red color and red precipitate appear in one of them. If heated long enough, the other sugar solutions should give a positive test also. However, record the order in which the sugars elicit positive tests and the extent of the reactions after twenty minutes.

E. Benedict's test

Pour 5 ml of Benedict's solution into each of eight test tubes labeled with the initial letter of glucose, arabinose, fructose, maltose, lactose, sucrose, ribose, and water. Add approximately 5 ml of water or 1% sugar solution as labeled on the appropriate tube. Heat the tubes in a boiling water bath for ten minutes. The formation of an orange to red precipitate indicates the presence of a reducing sugar. Record the relative extent of the reaction. Repeat the test on the plant material provided and record the results in Table 19–1.

F.   Phenylhydrazine test

Precaution: Phenylhydrazine is poisonous. Avoid contact with skin or breathing the vapors. Use a pipetting device or a burette.

Place 2 ml of each 1% sugar solution and distilled water into separate, labeled, test tubes. To each of the tubes add 5 ml of freshly prepared and filtered phenylhydrazine mixture and place them in a bath of boiling water for thirty minutes.

Allow the tubes to cool slowly and note when crystals form. Transfer a small quantity of the crystals to a slide and examine microscopically. You should note which sugars have identical crystals and which do not form osazones. Repeat the test on plant material provided. Record all results in Table 19–1.

TABLE 19–1   Results of carbohydrate tests*

Carbohydrate	Molisch	Barfoed	Bial	Seli-wanoff	Benedict	Phenyl-hydrazine
Glucose						
Arabinose						
Fructose						
Maltose						
Lactose						
Sucrose						
Ribose						
Control (H$_2$O)						
Plant material(s)						

* Indicate a negative test with a minus (–) and positive with plus marks (+). The relative extent of the positive tests may be indicated by increasing number of plus marks (relative basis). For the phenylhydrazine test, sketch the shape of the crystals formed.

G.   Enolization of sugars

In mildly alkaline solutions, a slow conversion of aldoses to ketoses and vice versa occurs until equilibrium is established. To observe this, make the following mixtures in test tubes:

Tube 1: 1 ml saturated Ba(OH)$_2$ plus 1 ml 1% glucose

Tube 2: 1 ml saturated Ba(OH)$_2$ plus 1 ml 1% fructose

Tube 3: 1 ml distilled water plus 1 ml 1% glucose

Tube 4: 1 ml distilled water plus 1 ml 1% fructose

Add two drops of preservative solution (1% thymol in toluene) to each tube and incubate for two hours. Then add 2 ml 0.1M HCl to each of tubes 1 and 2. If necessary, add additional HCl drop by drop until the contents are neutral or slightly acid to litmus. Add water to the other tubes (3 and 4) until all volumes are equal.

Place 1 ml of the contents of each tube into a fresh tube and 1 ml of Seliwanoff's reagent to each of the new tubes, and put them simultaneously into a boiling water bath. Note the order in which color appears in the various tubes. At the end of five minutes, compare the intensities of color in the different tubes.

Observations for enolization of sugars:

H. Application of the tests on plant material

Different kinds of plant materials such as sweet potato, apple, grapes, celery petioles, corn embryo, turnips, onions, and so on may be tested for carbohydrates. Juice may be extracted by grinding the material in a mortar and then filtering it. Use the filtrate equal to the volume of the various sugar solutions. The material may also be tested directly as fine shavings or from the mash obtained by grinding it in a mortar.

Results/ Conclusions

Record the results obtained for the various tests in Table 19–1. Present the results obtained for the enolization of sugars.

In concluding statements, consider the structure, the basis for classification, and the general biological roles of carbohydrates. In addition, interpret your results, indicate the basic reaction sequence involved in each test, and if it is a general or specific test for the detection of carbohydrates.

References

Bohinski, R. C. 1979. *Modern Concepts in Biochemistry*, 3rd edition. Boston: Allyn and Bacon.

Gander, J. E. 1976. Mono- and oligosaccharides. In Bonner, J. and J. E. Varner, (eds.), *Plant Biochemistry*, 3rd edition. New York: Academic Press.

Lehninger, A. L. 1982. *Principles of Biochemistry*. New York: Worth.

McGilvery, R. W., with G. Goldstein. 1979. *Biochemistry: A Functional Approach*. Philadelphia: W. B. Saunders Co.

Shriner, R. L., R. C. Fuson, and D. Y. Curtin. 1964. *The Systematic Identification of Organic Compounds*, 5th edition. New York: John Wiley and Sons, Inc.

White, E. H. 1970. *Chemical Background for the Biological Sciences*, 2nd edition. Englewood Cliffs, N. J.: Prentice-Hall, Inc.

White, A., P. Handler, E. L. Smith, R. L. Hill, and I. R. Lehman. 1978. *Principles of Biochemistry*, 6th edition. New York: McGraw-Hill.

The estimation of total soluble carbohydrate in cauliflower tissue

Introduction

During the course of their research, plant scientists are often required to determine the amount of carbohydrates in plant tissue. Carbohydrates exist in plant tissue as the polysaccharides starch and cellulose and a variety of substances classed as simple sugars (monosaccharides, disaccharides, for example) and oligosaccharides. Most of the non-polysaccharides are soluble in ethanol and are often termed *ethanol soluble carbohydrates*. This exercise deals with the extraction and quantitative determination of total ethanol soluble carbohydrates present in cauliflower tissue. The extracted carbohydrates react with anthrone reagent to produce a colored product that can be measured by standard colorimetric techniques.

The monosaccharides react readily with the sulfuric acid of the reagent mixture to form various furfuryl derivatives, which in turn react with the anthrone. The disaccharides, trisaccharides, and so on, however, are hydrolyzed by the sulfuric acid in the reagent mixture before conversion to various furfuryl derivatives and subsequent reaction with anthrone. The values obtained by colorimetric techniques are then compared to a standard curve determined from known amounts of glucose in order to estimate the total amounts of total soluble carbohydrates present in the extract.

Purpose

To determine the amount of total soluble carbohydrate in cauliflower tissue using anthrone reagent and standard colorimetric techniques.

Materials

cauliflower head

100 ml glucose solution (10 mg glucose/100 ml H$_2$O)

200 ml anthrone reagent (see Appendix II)

125 ml ethanol (80%)

18 test tubes

burette (50 ml)

water bath (500 ml beaker or equivalent)

tripod or ring stand and clamp

Bunsen burner

colorimeter or spectrophotometer and cuvettes

mortar and pestle

Buchner funnel and suction flask

filter paper (Whatman No. 1)

3 pipets (1 ml, graduated)

2 pipets (5 ml, graduated)

centrifuge

Procedure

A.   Standard curve of different glucose concentrations

Pipet the following amounts of a glucose solution (10 mg per 100 ml of water) into separate and appropriately labeled test tubes: 0.2, 0.4, 0.6, 0.8, and 1 ml. Adjust the total volume of each tube to 3 ml by the addition of distilled water. Prepare a reagent blank by adding 3 ml of water to another tube.

To all the test tubes, accurately add 6 ml of anthrone reagent from a burette and gently shake. Place all the tubes in a boiling water bath for three minutes, after which time they should be immersed in a cold water bath and allowed to cool.

After cooling, take readings using a suitable spectrophotometer or colorimeter set at 600 nm. Consult the references at the end of this exercise for background information concerning the principles of colorimetry and also question the instructor concerning the operation of the instrument to be used. Plot the optical density (or Klett units) against the amount of glucose in micrograms (see Figure 20–1).

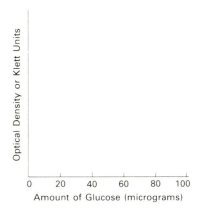

FIGURE 20–1   Standard curve of different glucose concentrations.

B. Estimation of soluble carbohydrate in cauliflower

Weigh exactly 1 g of cauliflower tissue and grind vigorously in a mortar with 20 ml of 80% ethanol. Filter the slurry with suction through Whatman No. 1 filter paper and save the filtrate.

Wash the residue off the filter paper with 80% ethanol and regrind it in 80% ethanol. Filter through paper again and repeat the grinding and filtration process one more time.

After the third grinding and filtration, discard the residue and combine all filtrates. Adjust the volume of the filtrate to 100 ml with 80% ethanol and centrifuge at 10,000 × g for fifteen minutes to obtain a clear supernatant.

Pipet 0.2, 0.4, and 0.5 ml aliquots of the supernatant into separate test tubes. Prepare three additional tubes in the same manner so as to have a duplicate for each aliquot.

Place all tubes in a boiling water bath and allow the ethanol to evaporate. Cool and add 3 ml of distilled water to each tube and mix. Also, prepare a reagent blank by adding 3 ml of water to another clean test tube. Add 6 ml of anthrone reagent to all tubes and repeat the test procedure outlined previously.

Results/ Conclusions

Employing the results from the standard curve of known glucose concentrations (Figure 20–1), determine the amount of soluble carbohydrate present in the cauliflower tissue (amount in micrograms per gram plant material extracted).

In concluding remarks, explain the operations of isolation and the reactions involved in testing carbohydrates with anthrone reagent. What does the term *soluble carbohydrate* mean, and what types of sugars are involved? Is glucose the only sugar present in the soluble carbohydrate fraction isolated from cauliflower?

Explain the principles of colorimetry including the Beer-Lambert Law, optical density, and percent transmission as they apply to the standard curve for different glucose concentrations.

Acknowledgment

This exercise was designed for class use by C. J. Pollard, Michigan State University, East Lansing.

References

Clark, J. M., Jr. (ed.). 1964. *Experimental Biochemistry*. San Francisco: W. H. Freeman and Co.

Gillam, A. and E. S. Stern. 1958. *Introduction to Electronic Absorption: Spectroscopy in Organic Chemistry*, 2nd edition. New York: St. Martin's Press.

Lehninger, A. L. 1982. *Principles of Biochemistry*. New York: Worth.

Miller, L. P. 1973. Mono- and oligosaccharides. In Miller, L. P. (ed.), *Phytochemistry*. New York: Van Nostrand Reinhold.

(Also see references following Exercise 19.)

EXERCISE 21

The separation of sugars by paper and thin-layer chromatography

Although gas chromatography is widely used for the separation and identification of sugars, the use of paper chromatography, even today, is valuable for the detection and separation of sugars present in plant extracts. Paper chromatography is not as precise as gas chromatography; yet it is often used in the absence of gas chromatographic equipment and has the advantage of being performed easily by an individual laboratory worker.

Paper chromatography as well as some types of thin-layer chromatography are based on the principle of partitioning the components to be separated between two liquid phases. In the first phase, the *stationary phase*, a thin water film saturates the cellulose of the paper. In the second, the *mobile phase*, usually an organic solvent flows over the cellulose and stationary water phase of the paper.

Initially, the sugars or components to be separated, by paper chromatography (pigments, amino acids, and other chemicals of plants) are spotted at intervals along the paper parallel to, but about an inch-and-a-half above, a straight edge (the bottom) of the paper. After the application of several microliters of extract, the spots are allowed to dry. After drying, the paper is placed into a chamber containing an organic solvent (usually water-saturated) so that the bottom of the paper is immersed in the solvent but the spotted extracts are above the solvent level. The solvent will move up the paper and through the extract spots. As the organic solvent continues to flow, the sugars will also move by partitioning between the mobile and stationary phases (that is, by entering the mobile phase, then into the stationary phase, back into the mobile phase, and so on). The rate that the sugars move upward is largely determined by their relative affinity for the two phases, or *partition ratio*. A relatively slow-moving sugar has greater affinity for the stationary phase, while for the faster-moving molecules the converse is true.

When migration of the mobile phase is stopped (usually when the solvent front is close to the top), the paper is removed from the chamber. The solvent front is marked with a pencil and the paper is hung to dry. After drying, the paper is sprayed with *p*-anisidine (for carbohydrates) or another suitable reactant and heated if necessary for the formation of colored products. The distance of the spots from the origin as compared with known standards often gives an indication of the sugar present. Also, the ratio of distance of the sugar from its origin (its point of application) in relation to the distance moved by the solvent (Rf value) will help to identify a sugar. For a better understanding of this process, consult with your instructor or read one of the standard texts on chromatography.

In chromatography, the solvent flows from the bottom of the paper upward; this is termed *ascending paper chromatography*. Another method of developing chromatograms involves the flow of solvent downward; the paper is oriented in a chromatographic chamber in such a fashion that the origin is still oriented close to the solvent reservoir. Chromatography with downward flow of the mobile phase is referred to as *descending chromatography*.

Essentially the same principles described above relate to the separation of sugars by thin-layer chromatography. In this procedure, however, the stationary phase is often a thin layer of an inert substance on glass or other support. As with paper, solvent migration may be ascending or descending.

Purpose

To illustrate the basic techniques used in the separation of sugars by paper and thin-layer chromatography.

Materials

unknowns (use any of the following): 2 to 3 unknown solutions, each containing two different sugars at 1% (w/v) concentration; fluid from nectaries of *Poinsettia* or *Asclepsis* flowers; 85% ethanolic extracts from cauliflower or other plant sources

8 separate 1% (w/v) sugar solutions of glucose, fructose, rhamnose, sucrose, arabinose, ribose, and galactose

300 ml Solvent I (*t*-butanol–glacial acetic acid–water, 3:1:1)

300 ml Solvent II (phenol–water, 4:1)

p-anisidine spray reagent (see Appendix II)

100 ml Solvent III (ethyl acetate–isopropanol–water–pyridine, 26:14:7:2)

2 Pyrex battery jars (preferably 10 by 18 in.)

2 ground glass lids

stopcock grease

oven (set at 80°C)

atomizer and bottle (or other spraying device)

1 large sheet chromatography paper (Whatman No. 1)

meter stick, ruler

scissors

micropipets

thin-layer development chambers with lids

Whatman No. 1 chromatography paper to line the development chambers

thin-layer silica gel G plate

Procedure

A.   Preparation of chromatogram chambers

The chromatograms should be run in 10 by 12 in. Pyrex battery jars fitted with ground glass plates to serve as lids (if not available, large-size pickle jars with screw tops may be used). To ensure that the jars are airtight, the rims should be greased with stopcock grease or vaseline, but do not use an excessive amount.

Two jars will be needed: one for Solvent I (*t*-butanol–glacial acetic acid–water, 3:1:1) and one for Solvent II (phenol–water, 4:1). Each jar should contain $1/2$ in. of the appropriate solvent as measured through the side of the jar. Cover the jars and allow the internal atmosphere to equilibrate for several hours before use.

B.   Preparation and development of the chromatograms

Note: See Figure 21–1. Cut a large sheet of Whatman No. 1 chromatography paper into halves along the short axis, using care to avoid contact with fingers or other surfaces that might contribute interfering substances. If the chromatogram chambers to be used are smaller than those indicated, the chromatography sheet may be cut into quarters and used with other minor alterations in the procedure.

Lay two cut halves of paper on a clean surface with the long axis parallel to the table edge. Draw a pencil line 2 in. above the edge of one of the long sides. Then make a series of pencil dots at intervals of 1 in. along the line and label these glucose, fructose, rhamnose, unknown, sucrose, arabinose, ribose, unknown, rhamnose, xylose, and galactose (Figure 21–1). The two papers should be duplicates but will go into separate solvent systems.

Arrange the sheets to overlap a support such as a glass tube or meter stick in such a way that the pencil spots are raised above the surface of the table.

With a micropipet, apply intermittently 3 μl of sugar solution (about 30 micrograms of sugar) on the appropriate spot. Make sure the spot is allowed to dry between applications. Apply the unknown(s) indicated in the Materials section or provided by the instructor in the same manner.

When the spots are dry, staple the sheet in such a way as to form a cylinder with the row of spots near the base. Use enough staples to keep the cylinder upright in the battery jar containing the solvent. For convenience, the chromatograms may be allowed to develop overnight, preferably from 11 P.M. to 8 or 10 A.M. After placing each paper in its respective chamber, cover the jars and leave the chromatograms until the solvent front approaches the upper edge of the paper. Remember, this should require about ten to twelve hours and plans should be made accordingly.

After development, remove the paper cylinders, mark the solvent front, and hang the papers to dry in their original orientation.

When the papers are thoroughly dry, wear disposable gloves and spray the papers with *p*-anisidine reagent (in a fume hood) and place them in an oven at 80°C for three minutes. Circle the resulting spots with a pencil and note the colors.

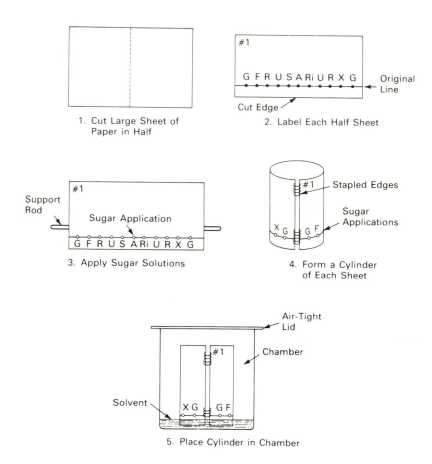

1. Cut Large Sheet of
   Paper in Half

2. Label Each Half Sheet

3. Apply Sugar Solutions

4. Form a Cylinder
   of Each Sheet

5. Place Cylinder in Chamber

FIGURE 21-1   Paper chromatography of sugars.

C. Thin-layer chromatography of sugars

On a line 3 cm from the bottom of a silica gel G plate provided by the instructor, mark seven 3-mm circles equidistant from each other and the sides of the
plate. Using a micropipet, apply small amounts of the unknown sugar solution to
one circle. Also spot known solutions of glucose, fructose, sucrose, arabinose, ribose, and galactose on separate circles and label. After the spots dry, place the
plate in a small chamber containing Solvent III. The level of the solvent should be
2 cm below the line of spots. Line the chamber with paper thoroughly wetted with
the solvent, it is most important that the atmosphere be saturated with the solvent
before use. Allow the plate to develop in the chamber. Before the solvent moves
over the end of the plate (which can take a matter of minutes), remove the plate
(use disposable gloves) and allow the solvent to evaporate in a fume hood. Draw a
pencil line along the solvent front. After the plates are thoroughly dry, spray with

the p-anisidine reagent in a fume hood and place in an oven at 80°C for three minutes. Circle the spots and note the colors. Calculate the Rf values of the known and the unknown solutions.

Results/ Conclusions

Calculate the Rf and Rrh values for each spot. The Rf value is obtained by determining the ratio of the distance the solute (sugar) moved divided by the distance the solvent moved (from the point of origin of the solute). The Rrh value is equal to the ratio of the movement of sugar(s) to the movement of rhamnose from the origin.

Record the Rf and Rrh values obtained for the respective sugars, as well as the color of the spots, in Table 21–1. Some aid is provided by the color. For example, aldohexoses should yield greenish-brown; aldopentose, yellow; uronic acids, red; methyl pentose, green; and deoxy sugars, light brown. Disaccharides usually respond with brownish tones. By comparing the color of the spots and the Rf and Rrh of the unknown to the standard, you can make an educated estimate as to whether or not the unknown contains some of the known sugars.

TABLE 21–1   Results of ascending paper chromatography of sugar standards and unknowns

Compound	Solvent I			Solvent II		
	Rf	Rrh	Color	Rf	Rrh	Color
Glucose						
Fructose						
Rhamnose						
Unknown						
Sucrose						
Arabinose						
Ribose						
Unknown						
Rhamnose						
Xylose						
Galactose						

In concluding remarks, consider some of the principles involved in chromatography with special emphasis placed on partitioning, surface adsorption, and ion-exchange. Define such terms as polarity, partition coefficient, the mobile origin phase, and the stationary water phase.

If Part C was carried out, list some of the advantages of thin-layer chromatography and also some of the problems with the procedure. Why are smaller chambers more important for use in thin-layer chromatography than in paper chromatography?

Acknowledgment   This exercise was designed for class use by C. W. Hagen, Jr., Indiana University, Bloomington.

References

Abbott, D. and R. S. Andrews. 1965. *An Introduction to Chromatography*. London: Longman, Limited.

Bobbitt, J. M., A. E. Schwarting, and R. J. Gritter. 1968. *Introduction to Chromatography*. New York: Reinhold Publishing Corp.

Lederer, E. and M. Lederer. 1957. *Chromatography. A Review of Principles and Applications*, 2nd edition. New York: American Elsevier Co.

Macek, K. (ed.). 1972. *Pharmaceutical Applications of Thin Layer and Paper Chromatography*. Amsterdam: Elsevier.

Touchstone, J. C. and M. F. Dobbins. 1978. *Practice of Thin Layer Chromatography*. New York: John Wiley and Sons.

Zweig, G. and J. Sherma (eds.). 1972. *Handbook of Chromatography*. Cleveland: Chemical Rubber.

EXERCISE 22

Isolation and properties of starch

Introduction

Much of the sugar produced by photosynthesis is ultimately converted to starch, a polysaccharide synthesized in plastids as starch grains. This common storage product of plants may be found in seeds, tubers, bulbs, and other storage organs. Starch grains often differ in size and shape and are large enough to be observed microscopically.

Starch is actually composed of two polysaccharides, amylose and amylopectin. Both polysaccharides are polymers of α-D-glucose, which is produced by hydrolysis. Amylose is a straight-chain polymer of glucose molecules, and amylopectin is branched. In amylose, the straight chain is maintained by α-1,4-links in addition to α-1,4-links between glucose molecules. In contrast, amylopectin is characterized by the presence of α-1,6-links, and sometimes α-1,3-links in addition to α-1,4-links. The branching of the amylopectin molecule is therefore due to the number of α-1,6-links present and, in some polymers, the α-1,3-linkages.

It is interesting to note that, due to the structural differences of amylose and amylopectin, the former polysaccharide is responsible for the blue-black color that occurs when iodine is added to starch. This may be due to the coiled nature of amylose. Conversely, amylopectin gives a red to purple color when stained with iodine.

Starch is relatively easy to isolate from plant tissue containing abundant starch grains. By homogenization and/or grinding the cells, cellular membrane may be fragmented and the starch grains released. The "heavy" starch grains may be precipitated by low-speed centrifugation and washed, followed by suspension of the starch in boiling water.

It is well known that in plant tissues containing starch there are also present hydrolyzing enzymes. The two most common enzymes are α- and β-amylases, which are of primary importance in the degradation of starch in plants. In fact, they represent the best means for the mobilization of carbohydrate reserves in plants. However, their mode of action is quite different even though they both attack α-1,4-linkages only. β-amylase catalyzes the removal of maltose units one by one from the non-reducing end of a chain of glucose units. If the amylose molecule is composed of an odd number of glucose units, hydrolysis with β-amylase results in the formation of maltose and maltotriose molecules. The maltotriose molecule represents the three glucose units of the reducing end of the amylose molecule. If the molecule is amylopectin, then β-amylase can start at the non-reducing end of each branch and successively remove maltose units to within two glucose units of the α-1,6-linkages. In contrast, α-amylase attacks any α-1,-4-linkage at random on the starch molecule. That is, α-amylase may hydrolyze α-1,4-links at either end or in the middle of the molecule. If a branched chain is attacked, all α-1,4-links to within three units of the α-1,6-link can be hydrolyzed. Thus the products of α-amylase activity on starch are a variety of oligosaccharides or dextrins.

Another important enzyme involved in the synthesis as well as the degradation of starch in plants is starch phosphorylase. Regarding synthesis, it has been found that this enzyme will catalyze the formation of a polymer of glucose if glucose-1-phosphate and a primer molecule (consisting of three to an optimal number of twenty glucose residues in α-1,4-links) are present. The glucose of glucose-1-phosphate is added to the non-reducing end of the primer molecule to form an α-1,4-linkage. Thus, by the repeated addition of glucose to the non-reducing end of the primer molecule, the construction of a straight-chain amylose molecule results.

During the degradation of starch, starch phosphorylase in the presence of inorganic phosphorus catalyzes the phosphorolytic cleavage of the α-1,4-link of amylose to form glucose-1-phosphate molecules. This process is termed *phosphorolysis*, and not hydrolysis, because it involves phosphoric acid instead of water. High concentrations of inorganic phosphate and high pH (basic pH) favor phosphorolysis. Low pH (acid) and lower concentrations of inorganic phosphorus favor starch synthesis.

Purpose

To isolate and study the properties of starch and demonstrate amylase and phosphorylase activity.

Materials

10 corn grains, soaked in running tap water for 24 hours at 25°C

8 barley seeds, soaked initially in 50% (v/v) sulfuric acid for one-half hour and then soaked for 24 hours in running tap water (25 to 30°C)

4 petri dishes, containing 0.15% (w/v) starch solution solidified with 1.0% (w/v) agar

4 petri dishes (10 cm) containing a solidified preparation of 1.5% (w/v) agar only

4 petri dishes (10 cm) containing 0.5% (w/v) glucose-1-phosphate solidified with 1.5% (w/v) agar

iodine (I₂KI) reagent (see Appendix II)

phosphate-citrate buffer, pH 5.6 (see Appendix II)

0.5% (w/v) starch solution (see
  Appendix II)
Benedict's reagent (see Appendix II)
balance
Waring blender
cheesecloth
centrifuge
conical centrifuge tubes (heavy-
  duty)

10 test tubes
mortar and pestle
spot plate
dissecting equipment
microscope
microscope slides
coverslips

Procedure

A. Isolation of starch from potatoes

Homogenize 50 g of peeled, sliced white potato with 100 ml of distilled water in a Waring blender for two minutes. Filter the slurry through four layers of cheesecloth and allow the filtrate solution to stand for three minutes. Discard the residue left on the cheesecloth.

Decant the filtrate carefully and save the sediment of starch. Transfer the starch to 12 ml heavy-duty conical centrifuge tubes with a minimum amount of water and centrifuge at approximately 2400 rpm for two minutes. Decant, and discard the supernatant solution.

Wash the starch by resuspending in water (add enough water so that the centrifuge tube is $3/4$ full) and recentrifuge at 2400 rpm for two minutes. Repeat the washing procedure and then resuspend one-half of the starch in 1 to 2 ml of water. Cap the centrifuge tube (using a suitable cover or thumb) and vigorously shake.

Pour the thoroughly shaken material into 100 ml of boiling water in a 250 ml beaker. Cool the solution by placing the beaker into a cold water bath. While the solution is cooling, add 100 ml of distilled water and stir with a glass rod. The solution may be used when it reaches room temperature.

B. Properties of starch (iodine test)

Place two drops of the extracted starch solution in a test tube and add 3 ml of distilled water. Add a drop of iodine solution to the mixture and note the color. Place several drops of the colored mixture on a microscope slide and observe with a microscope. Place the remainder of the solution in a water bath and heat until a visible change of color takes place. Remove the tube from the water bath and allow it to cool in a bath of cold water. Note the results. Repeat with a known solution of soluble starch (5%).

Observations:

C. Effect of amylase from germinating barley seeds on starch

Gently grind eight germinating barley seeds, initially soaked in 50% (v/v) H_2SO_4 and then in running tap water for twenty-four hours at 25°C, in a mortar with 4 ml of phosphate-citrate buffer (pH 5.6). Carefully decant the liquid into a 12 ml conical tube and centrifuge for two minutes at 2400 rpm. Into four test tubes, introduce 0.5 ml of the supernatant solution. Heat two of the test tubes in a boiling water bath for four minutes and allow to cool. If any liquid is lost, replace it with an equal amount of buffer.

Introduce 0.5 ml of a 0.5% soluble starch solution (w/v) or the starch solution from potato to each of the four test tubes (two heated and two not heated), and test the contents of each at two-, five-, fifteen-, and twenty-minute intervals for starch. Two drops of the solution and a small drop of iodine in a spot plate should be adequate. Also, test two drops of each solution with Benedict's reagent.

Observations:

D. Hydrolysis of starch by amylase from corn scutellum

Note: Soak corn grains in running tap water for twenty-four hours prior to the exercise.

Prepare 1% plain agar containing 0.15% soluble starch by adding the agar and starch together in water and boiling until the agar is melted. Then pour not more than 15 ml of the solution of starch and agar into a petri dish. Sterilization is not necessary for a short-term experiment unless the petri dishes are prepared some time prior to the class meeting. Allow the solutions to solidify at room temperature.

Remove the endosperm from corn grains and place three embryos in each petri dish containing the solidified starch agar, with the scutellum in contact with the agar. Be sure the scutellum is used and not endosperm.

Incubate in a warm dark place for twenty-four hours. Remove the embryos and test for starch by pouring a standard I_2KI solution over the agar in the dish. Then quickly rinse with distilled water.

Observations:

E.   Conversion of glucose-1-phosphate to starch by potato phosphorylase

Prepare a solution of 1.5% agar and another solution of agar containing 0.5% glucose-1-phosphate. Boil and pour about 15 ml of each solution into appropriately labeled petri dishes and allow to solidify.

Slice a potato and homogenize in a Waring blender. Then carefully filter the homogenate through cheesecloth in order to remove the starch grains. Test an aliquot of the filtrate with I_2KI solutions. Repeat the filtration process until filtrate aliquots give a negative test with I_2KI. Careful and repeated filtering should remove the starch grains. Low-speed centrifugation may speed up their removal.

Place drops of the crude potato extract on the surface of the two different agar preparations, and test at different time intervals for starch (drop I_2KI on one spot at a time). Compare the results of the two agar preparations with time. Since the activities of the enzymes may vary greatly, different groups of students should, at the discretion of the instructor, choose different time intervals between drops of I_2KI.

Observations:

Results/
Conclusions

Present your results for each part as concise statements and, when applicable, design tables or other illustrations of the results. In concluding statements, consider the following:

1. Where starch is found in plant cells and in what form.

2. The properties of starch with respect to isolation, solubility in water, and staining with iodine reagent.

3. The change in color when the starch iodine complex is heated and then cooled.

4. The chemical nature of starch as compared to structural carbohydrates such as cellulose.

5. The chemical differences between amylose and amylopectin.

6. The action of phosphorylase in starch synthesis.

7. The action of amylase in starch degradation and pertinent details of the reaction involved.

References

Bohinski, R. C. 1979. *Modern Concepts in Biochemistry*, 3rd edition. Boston: Allyn and Bacon.

Duffus, C. M. and J. H. Duffus. 1984. *Carbohydrate Metabolism in Plants*. London: Longman, Limited.

Geiger, D. R. 1979. Control of partitioning and export of carbon in leaves of higher plants. *Bot. Gaz.* 140: 241–248.

Liu, T. T. Y. and J. C. Shannon. 1981. A nonaqueous procedure for isolating starch granules with associated metabolites from maize (*Zea mays* L.) endosperm. *Plant Physiol.* 67: 518–524.

Matheson, N. K. and R. H. Richardson. 1976. Starch phosphorylase enzymes in developing and germinating pea seeds. *Phytochemistry* 15: 887–892.

Preiss, J. 1982. Regulation of the biosynthesis and degradation of starch. *Ann. Rev. Plant Physiol.* 33: 431–454.

Shannon, J. C. 1978. Physiological factors affecting starch accumulation in corn kernels. Chicago: *Thirty-Third Ann. Corn and Sorghum Res. Conf.*

EXERCISE 23

Paper chromatography of nucleic acid bases and the ultraviolet absorption spectrum of ribonucleic acid or deoxyribonucleic acid

Introduction

As illustrated in Exercise 20, various plant components may be separated by paper chromatographic techniques. The following exercise is designed to illustrate a method for the separation of the common bases that are components of RNA and DNA. When appropriately applied to paper, purine and pyrimidine bases may be separated by the use of various solvent systems. The developing solvent system used in this exercise consists of a mixture of isopropanol, HCl, and water. After the chromatograms are developed for the prescribed tissue (approximately fifteen hours), they are removed from the chamber and allowed to dry. The bases may then be detected under ultraviolet light. The solvent part and ultraviolet absorbing spots (representing the bases) may be marked for calculation of the Rf values and assessment of relative migration rates of the individual bases.

The absorption spectrum of RNA and DNA can be determined over a wavelength range of 220 to 300 nm with the use of a suitable spectrophotometer. Before using the spectrophotometer, be sure to consult with the instructor for directions as to its proper use and operation.

Purpose

To chromatograph the major purine and pyrimidine bases present in nucleic acids and to determine the ultraviolet absorption spectrum of ribonucleic acid or deoxyribonucleic acid.

Materials

separate aqueous solutions of adenine, guanine, cytosine, uracil, and thymine (all in concentrations of 2 mg/ml)

chromatographic solvent (65 ml isopropanol, 16.7 ml 12N HCl, and final volume adjusted to 100 ml with H_2O)

0.1M tris HCl buffer, pH 7.6 (see Appendix II for preparation)

solutions of RNA or DNA (10 mg commercial nucleic acid in 500 ml 0.1M tris HCl buffer, pH 7.6)

1 sheet chromatography paper (Whatman No. 1)

descending chromatography apparatus

hood

darkened room (for studying chromatograms)

mineral lamp (short wavelength ultraviolet light)

meter stick

2 silica cuvettes

ultraviolet spectrophotometer

Procedure

A. Paper chromatography (descending) of nucleic acid bases

Chromatographic chambers

Prepare the descending chromatographic chamber by first placing an open beaker of solvent (65 ml isopropanol, 16.7 ml 12N HCl, and diluted up to 100 ml with water) on the bottom. Now fill the trough about three quarters full with the same solvent. Cover the chamber and allow the internal atmosphere to equilibrate before using (about one hour).

Chromatogram preparation

Cut a portion of Whatman No. 1 chromatography paper into strips that will accommodate the particular descending chromatography apparatus to be used. The length of the paper should allow for a distance of 35 cm for the solvent to run from the origin to the front.

Lay the paper down on a clean bench and draw the origin line so that it will be approximately 1 in. from the antisiphon bar fold when the paper is placed in the chromatographic chamber. Now draw circles (about ½ in. in diameter) 1½ in. apart along the origin line. Under each circle indicate the specific base to be applied and an unknown, if any. The analysis of a hydrolyzed plant extract is left to the discretion of the instructor.

Using a micropipet, apply between 10 and 25 ml of each base solution to the approximately labeled circle. Be sure to dry the spots after each application (use a stream of air or hair dryer).

Development of chromatograms

Initiate development of the chromatogram by immersing the end of the spotted paper closest to the origin in the chamber trough. In most chromatography systems, two sheets of paper can be developed simultaneously.

Allow the papers to develop at room temperature for fifteen hours or until the solvent has moved 35 cm from the origin.

After appropriate development, carefully remove the paper so as not to touch any surfaces and allow to dry in a hood. After thorough drying, observe the chromatogram in a darkened room under irradiation from an ultraviolet light (short wavelength) mineral lamp. Do not look directly at the ultraviolet source, and wear glasses.

Mark the solvent front and the dark, ultraviolet-absorbing spots with a pencil and calculate the Rf values by dividing the distance from the origin to the solvent front into the distance from the origin to the center of each spot.

Calculation of Rf values:

B. **Ultraviolet absorption spectrum of ribonucleic acid (RNA) or deoxyribonucleic acid (DNA)**

Dissolve 10 mg of RNA or DNA in 500 ml of 0.1M tris HCl buffer (pH 7.6). Pour a small amount of the solution into one of two silica cuvettes. For a reagent blank, fill the second cuvette only with buffer.

Place both cuvettes into an ultraviolet spectrophotometer and determine the absorption spectrum from 220 to 300 nm. Take and record optical density readings for the nucleic acid solution every 5 nm. Over the range indicated, at the points of maximum and minimum absorption, take additional readings at 1 nm intervals. Record the readings.

Optical density readings:

At the discretion of the instructor, an unknown amount of either RNA or DNA will be provided. Design a working curve of absorbance based on the known amount of the nucleic acids. Using the curve, calculate the amount of unknown material in the sample.

**Results/
Conclusions**

Present the Rf values obtained for each of the bases. Present a graph of the ultraviolet absorption spectrum for either DNA or RNA in which the optical density readings are plotted on the ordinate against the wavelengths of the abscissa.

References

Bedbrook, J. R. and R. Kolodner. 1979. The structure of chloroplast DNA. *Ann. Rev. Plant Physiol.* 30: 593–620.

Bohinski, R. C. 1979. *Modern Concepts in Biochemistry*, 3rd edition. Boston: Allyn and Bacon.

Cantoni, G. L. and D. R. David (eds.). 1966. *Procedures in Nucleic Acid Research*. New York: Harper and Row.

Flavell, R. 1980. The molecular characterization and organization of plant chromosomal DNA sequences. *Ann. Rev. Plant Physiol.* 31: 569–596.

Lehninger, A. L. 1982. *Principles of Biochemistry*. New York: Worth.

McGilvery, R. W., with G. Goldstein. 1979. *Biochemistry: A Functional Approach*. Philadelphia: W. B. Saunders Co.

Semenko, G. L. 1965. *Biochemistry of Nucleic Acid Metabolism in Higher Plants*. New York: Reinhold Publishing Corp.

Watson, J. D. and F. H. C. Crick. 1953. Molecular structure of nucleic acids. *Nature* 171: 737–740.

EXERCISE 24

Adenosine hydrolase

Introduction — Under appropriate conditions, adenosine hydrolase will catalyze the hydrolysis of the nucleoside adenosine to the free base adenine. This reaction is important in the turnover of nucleic acids and as one of the steps providing free adenine for the production of adenine-containing compounds.

A good source of the enzyme is cauliflower florets. As part of the following exercise, adenosine hydrolase is to be extracted from cauliflower florets and the resulting extract tested for activity. As in all enzyme assays, the substrate is provided in the reaction mixture with any necessary cofactors and buffer (at the appropriate pH) for optimum activity. At the end of the reaction time(s), the resulting product is measured. In this case, the amount of adenine formed is determined by paper chromatographic separation, followed by detection of adenine on chromatograms with an ultraviolet lamp. Since several reaction mixtures will be run for different times, several chromatograms will be developed, and the extent of the reaction will be evaluated by the gradual disappearance of adenosine and the increase of adenine with time. Be sure to read the directions carefully so that suitable reaction mixtures are prepared and evaluated to determine the relative enzyme activity over time.

Purpose — To study the enzymatic hydrolysis of adenosine to adenine in the presence of a specific nucleosidase extracted from cauliflower.

Materials

16 g cauliflower

50 ml 0.1M trishydroxymethy-laminomethane (tris) buffer adjusted to pH 7.4 with HCl

13 ml adenosine solution (2 mg/ml; made up in 0.1M tris buffer at pH 7.4)

200 ml boric acid solution (0.03M; adjusted to pH 8.4 with NaOH)

cheesecloth

cotton

funnel

4 capillary pipets (or medicine droppers drawn to a fine tip)

ice water bath

boiling water bath

2 chromatography paper sheets (Whatman No. 1, 28 × 23 cm)

2 chromatography chambers (to accommodate the paper when rolled in a cylinder)

short-wave ultraviolet light source (mineralight)

4 vials or small test tubes

Procedure

To complete the exercise in one class period, two or more students should work together.

A. Enzyme preparation

Note: All reagents and glassware used in the preparation of the enzyme should be ice-cold.

Grind 15 g of cauliflower florets in a cold mortar with 15 ml of cold tris buffer (pH 7.4). Squeeze the grindate through a double layer of cheesecloth and into a small pad of glass wool in a funnel. The resulting filtrate contains the enzyme and should be kept cold and used as indicated in the following experiment. In addition, prepare a small amount of inactivated enzyme by transferring 3 ml of the filtrate to a clean test tube. Place the tube in a boiling water bath for fifteen minutes. Remove the tube from the water bath and adjust the volume back to 3 ml with buffer if any of the liquid was lost during heating. This preparation contains inactivated enzyme and should be used where indicated (as boiled enzyme) in the experiment.

B. Assessment of enzyme activity on the basis of adenine production

Prepare a sheet of chromatography paper (Whatman No. 1) 28 by 23 cm. Orient the long edge of the paper parallel to the bench top edge and draw a line about 1 in. from the bottom and along the long axis of the paper.

Add the reaction mixture components to small glass containers according to the following outline (do not add the enzyme until you are ready to start the reactions):

1 ml tris buffer (pH 7.4) + 1 ml filtrate

1 ml adenosine (2 mg/ml) + 1 ml filtrate

1 ml adenosine (2 mg/ml) + 1 ml filtrate (boiled)

1 ml adenosine (2 mg/ml) + 1 ml buffer (pH 7.4)

Since the filtrate may be quite active it may be necessary to have several groups of students use different amounts of the filtrate (for example, 0.2, 0.4, 0.6, or 0.8 ml of filtrate) so that the conversion of adenosine to adenine with time can be illustrated. Accordingly, adjust the final volume of the reaction mixture with buffer.

Immediately after starting the reactions, shake the mixtures and apply a small amount of each as separate spots on the origin line of the chromatography paper. Be sure to start from the left-hand side (about ½ in. from the edge) and work to the right. Do not overlap the individual reaction mixture applications.

Label the set of four applications as "0 time." Thirty minutes later, make additional applications for each of the reaction mixtures and to the right of the first set of spots. Label this set of applications "30 minutes."

Over the next hour and a half, repeat the application procedure at thirty-minute intervals so that by the end of the total two-hour reaction time, you will have made a total of five sets of applications, each consisting of four spots representing each reaction mixture. Use an additional sheet of chromatography paper if necessary. Then make a cylinder of the paper and develop the chromatogram in a chamber with 0.03M boric acid (pH 8.4) serving as the ascending solvent. Additional time outside of class is required for chromatogram development and observations. Therefore, plans should be made accordingly.

The solvent will travel to the top of the cylinder in slightly over one hour. In addition, adenosine will travel closer to the solvent front than does adenine. After drying the chromatograms, adenosine and adenine can be detected with an ultra-violet lamp (*Caution!*). Note the relative intensities of each compound with time. Calculate the Rf for each ultraviolet-quenching compound and use plus marks on a relative basis for illustrating the results.

Observations:

Results/ Conclusions

Present the Rf values for adenosine and adenine for the specific solvent system used. Indicate the relative amounts of adenine and adenosine detected on the chromatograms at the different time intervals studied.

In concluding statements, interpret the results and present an equation that illustrates the reaction studied. The answers to the following questions should also be included:

1. What is the effect of time on enzyme activity?

2. What other factors influence enzyme activity?

3. Is the extracted nucleosidase specific with respect to the nucleoside used in the reaction mixtures?

4. What is the possible significance of adenosine hydrolase in the intact plant?

Acknowledgment

This exercise was adapted for class use by C. O. Miller, Indiana University, Bloomington.

References

Heppel, L. A. and R. J. Hilmore. 1952. Phosphorylases and hydrolysis of purine ribosides by enzymes from yeast. *Biol. Chem.* 198: 683–694.

Mazelis, M. 1959. Enzymatic degradation of adenosine triphosphate to adenine by cabbage leaf preparations. *Plant Physiol.* 43: 153–158.

Mazelis, M. and R. K. Creveling. 1963. An adenosine hydrolase from brussels sprouts. *J. Biol. Chem.* 238: 3358–3361.

(Also see the references following Exercise 23.)

EXERCISE 25

Proteins

Introduction

The need to measure the quantities of different chemical substances invariably arises in the course of physiological studies. Often the substances are present in microgram or milligram amounts as well as in complex mixtures. The investigator must consider several criteria including sensitivity, specifity, ease, and cost problems when selecting an appropriate assay for a desired substance. Measuring protein quantity presents an unusually complete example of the considerations and problems that often confront the investigator. Because each protein has a unique primary sequence and therefore amino acid composition, assays based on amino acid side chain reactivities or R-group ultraviolet light absorption will give different responses for different proteins. Alternatively, assays based on general protein properties—that is, the peptide bond—often lack sensitivity and require high protein concentrations. Finally, many preparations of proteins are designed to preserve enzymatic activity and therefore contain added components (buffer, reducing agents, and EDTA) that might interfere with the assay. It is not surprising that a large number of assays have been developed to measure proteins.

The investigator can prevent those problems by knowing the basis of the assay, practicing repeatable protocols, and using a defined, reproducible protein standard.

Purpose

To demonstrate the use of standard tests for the detection of proteins in plant materials, to establish standard curves for protein determination, and to compare different procedures for measuring proteins.

111

Materials

80 g corn grains, soaked in running tap water for 24 hours

30 ml egg albumen solution (10 mg/ml)

0.5N NaOH

mortar and pestle

filter paper

pipets (graduated: 1, 5, and 10 ml)

Buchner funnel and side-arm flask

water bath

test tubes

Vis-UV-spectrophotometer

cuvettes

Biuret reagent (1.5 gm $CuSO_4$ $5H_2O$ and 6 gm Na-K tartrate dissolved in 500 ml of H_2O, added to 300 ml 10% NaOH and diluted to 1 liter)

Reagent 1 (1 ml of 2% $CuSO_4$ mixed with 1 ml 4% Na-K tartrate diluted to 50 ml with 3% $NaCO_3$ in 0.1N NaOH)

Reagent 2 (Folin-Ciocalteu's phenol reagent)

2N phosphomolybodate-phosphotungstate diluted 1:2 with H_2O just before use

Procedure

A. Preparation of extract from germinating corn grains

Soak corn grains overnight under running tap water. Grind approximately 80 g of soaked kernels with a mortar and pestle in about 30 ml of water. Filter with a Buchner funnel and wash the residue with 15 ml of water. Combine the filtrates and use in the tests outlined below.

B. Protein tests

Each of the following tests should be performed.

Biuret procedure

The Biuret test involves the production of a pink color by proteins in an alkaline solution of cupric ions. The color is believed to result from a complex between the peptide backbone of the protein and cupric ions.

1.   Experiment: To six test tubes add 0, 0.1, 0.2, 0.3, 0.6, and 1 ml of egg albumen in 10 mg/ml. Adjust volume to 1 ml with H_2O. Similarly, add 0.1, 0.5, and 1 ml of the corn extract to three tubes and bring the total volumes to 1 ml. Add 4 ml of Biuret reagent to each tube, mix thoroughly, and incubate the reactions for thirty minutes at room temperature. Read the absorbance of all tubes against the zero protein blank at 540 nm. Plot a standard curve using egg albumen and estimate the protein concentrations of the corn extract.

Folin-Ciocalteu procedure

The Folin-Ciocalteu reaction develops color due to the copper-catalysed reduction of phosphomolybdic-phosphotungstic acid by primarily tyrosine, tryptophan, and cysteine. Other reducing compounds in the protein mixture could also react.

2. Experiment: Dilute an aliquot of the egg albumen and corn extract 1:25 with 0.5N NaOH. To five test tubes add 0, 0.1, 0.2, 0.3, and 0.5 ml of diluted egg albumen, bringing the total volume to 1 ml with 0.5N NaOH. To each of three tubes add 0.1, 0.3, and 0.5 ml of the corn extract and adjust the total volume to 1 ml with 0.5N NaOH. Add 5 ml of reagent, 1 ml to each tube, mix immediately, and incubate at room temperature for ten minutes. Add 0.5 ml of Reagent 2, mix, incubate thirty minutes, and read the absorbance at 660 nm against the zero protein blank. Plot a standard curve for egg albumen and estimate the protein concentration of the extract.

UV absorption

Proteins absorb in the region of 280 nm in the ultraviolet portion of the spectrum, primarily because of the aromatic amino acids tryptophan and tyrosine. Although not as sensitive as the colorimetric procedures used above, direct photometry is nondestructive and easy.

3. Experiment: Measure the absorption spectrum of egg albumen and the corn extract at 5 nm intervals from 350 to 230 nm. Start at the longer wavelengths because they are of the lowest energy and least destructive. Plot the absorption spectrum and determine the absorption per mg egg albumen at the peak wavelength. Estimate the protein concentration of the corn extract from this value.

Results/ Conclusions

In concluding statements, consider the chemistry of proteins with respect to peptide bonding, structure (primary, secondary, and tertiary), and the biological roles of proteins. Compare the different tests performed for sensitivity, specificity, and ease. What other chemicals, natural or buffer components, possibly might also react in the three tests?

Acknowledgment

The experiments in this exercise were prepared by Dr. Charles Boyer, Department of Horticulture, The Pennsylvania State University.

References

Ellis, R. J. 1981. Chloroplast proteins: Synthesis, transport, and assembly. *Ann. Rev. Plant Physiol.* 32: 111–137.

Fincher, G. B., B. A. Stone, and A. E. Clarke. 1983. Arbinogalactan-proteins: Structure, biosynthesis, and function. *Ann. Rev. Plant Physiol.* 34: 47–70.

Lamport, D. T. A. 1980. Structure and function of plant glycoproteins. In Preiss, J. (ed.), *Biochemistry of Plants*, Vol. 3. New York: Academic Press, pp. 501–541.

Lehninger, A. L. 1982. *Principles of Biochemistry*. New York: Worth.

Stryer, L. 1981. *Biochemistry*, 2nd edition. San Francisco: W. H. Freeman and Co.

EXERCISE 26

Separation of amino acids by two-dimensional paper chromatography

Introduction

Amino acids present in plant extracts may be separated by conventional paper chromatographic techniques, although current techniques employ the use of gas chromatography, column chromatography, or an amino acid analyzer.

The following exercise provides the basic techniques for separating and detecting amino acids by two-dimensional paper chromatography. Since the fundamental principles of paper chromatography have been previously discussed (see Exercise 21), only those procedures relating to two-dimensional paper chromatography will be considered here.

An amino acid extract is applied with a micropipet as a spot on the bottom edge of chromatographic paper. The material is applied so that the spot is about two inches from the bottom and approximately the same distance from the left-hand edge. Small quantities of extract are added intermittently, with time allowed for drying between each application.

After the material is applied and dry, the paper is placed into a chamber containing solvent for ascending or descending chromatography. The chamber is covered and the chromatogram is allowed to develop until the solvent front approaches the top of the paper (about two to three inches from the top). The chromatogram is then dried. At this point, some of the amino acids should have separated from the origin to the top as an array in line. The dried chromatogram is then placed into another chamber containing a different solvent so that the original left-hand edge of the paper is now immersed. As the second solvent migrates, it will move over the line of previously separated amino acids (now the origin). When the solvent is close to the top (ascending) or bottom (descending), the chromatogram is removed and dried. After drying, the chromatograms are sprayed (for amino acids, ninhydrin reagent is generally used) and heated. During the heating process, distinct colored spots (dark blue, light blue, sometimes pink) will appear on the paper. These spots represent different amino acids, the migration of which (through Rf determinations and use of standards) will provide some

114

indication of the amino acids present. This method of separation of amino acids by two-dimensional chromatography is often referred to as a "fingerprinting" technique and, as you will observe, is valuable in the determination of various amino acids present in different plant extracts.

Purpose

To illustrate the techniques involved in the separation of amino acids by two-dimensional paper chromatography.

Materials

30 g peeled white potato	filter paper (Whatman No. 1)
10 ml amino acid mixture consisting of alanine, tyrosine, serine, and glutamic acid (concentrations of each amino acid — 1 mg/ml of 0.5N acetic acid)	Buchner funnel and side-arm flask
	tared weighing dish
	1 large sheet chromatography paper (Whatman No. 1)
100 ml ethanol (80%, v/v)	chromatography jars (10 by 18 in.) and lids
350 ml phenol solvent (water-saturated phenol)	2 micropipets
350 ml butanol solvent (*n*-butanol–glacial acetic acid–water, 3:1:1)	stapler, staples
	oven (set at 90°C)
ninhydrin spray reagent (see Appendix II)	atomizer and bottle (or other suitable spraying device)

Procedure

A. Amino acid extract

Blend approximately 30 g of peeled white potato with 100 ml of 80% ethanol. Filter through a Buchner funnel and evaporate the filtrate to dryness under reduced pressure at room temperature. Perform the final steps of the evaporation in a tared dish so that an estimate of the amount of solid may be obtained. Just before use, redissolve the residue in sufficient distilled water to give about 30 mg of solid per 1 ml water.

B. Paper chromatography chambers

Paper chromatograms may be developed in 10 by 18 in. Pyrex battery jars fitted with ground glass plates to serve as lids. The rims of the jars may be greased with stopcock grease, but do not use an excessive amount.

Four jars will be needed: two for phenol solvent (*Caution! Phenol is caustic and toxic!*) and two for the *n*-butanol–acetic acid–water mixture. The solvent levels in each jar should measure about ½ in. from the bottom.

After introducing the solvent, immediately cover the jars and allow the internal atmosphere to equilibrate for several hours.

C. Preparation and development of the chromatograms

Cut a sheet of Whatman No. 1 filter paper into quarters, using care to avoid contact of the paper with fingers or other surfaces that might contribute interfering impurities.

Lay two quarter sheets on a clean surface with the long axis parallel to the table edge. With a pencil, make a light dot 2 in. from the left-hand edge and $1\frac{1}{2}$ in. from the bottom on each sheet. On the lower edge of each sheet, pencil the word "phenol" and number the sheets "1" and "2."

Arrange the sheets to overlap a support such as a glass plate in such a way that the pencil spots are raised above the surface. Using a micropipet, deposit 3 μl of the potato extract on the pencil dot of sheet 1, but not all at once. The extract should be applied slowly and intermittently so that the final spot has a diameter of about 1 cm. On sheet 2, deposit an equal volume of the known amino acid mixture provided by the instructor.

When the spots are dry, form a single cylinder of each sheet with the edges marked "phenol" forming the base. Use enough staples to keep the cylinder rigid, but leave a space between the two edges of the paper. Chromatogram development in this solvent requires ten to twelve hours; therefore plan accordingly. Stand the cylinder upright in the battery jar containing the phenol solvent. Cover the jar and leave until the solvent front approaches the upper edge of the paper. At this point, remove the cylinder, mark the position of the solvent front, and hang up to dry in its original orientation.

When dry, unstaple the cylinder and trim $\frac{1}{2}$ in. off the stapled edge. The amino acids should now be distributed as a line along the left-hand side of each paper. Rotate the papers to bring this line to the bottom, then staple to form a new cylinder and place in the n-butanol–acetic acid–water mixture. (Development takes ten to twelve hours.) Just before the solvent reaches the top of the paper, remove the cylinder, mark the solvent front, and dry.

When thoroughly dry, the paper should be sprayed uniformly and lightly with ninhydrin reagent (0.2% ninhydrin in water-saturated n-butanol). Do this in a hood using a bulb atomizer. Heat the sprayed paper at 90°C for five minutes, then outline the spots with a soft pencil and note the colors. Most of the amino acids give purple spots, but phenylalanine, tyrosine, and aspartic acid give blue colors; tryptophan, olive brown; asparagine, cystine, and cysteine, brown; and proline, yellow.

Results/
Conclusions

Calculate the Rf values of the important spots by determining the ratio of the travel of the solute to the travel of the solvent from point of origin of the solute.

From the colors of the spots and from their Rf values (see published values) and relative positions on the paper, it should be possible to identify tentatively several of the amino acids present in the potato extract.

In concluding remarks, consider the principles involved in paper chromatography.

Acknowledgment

This exercise was designed for class use by C. W. Hagen, Jr., Indiana University, Bloomington.

References

Bobbitt, J. M., A. E. Schwarting, and R. J. Gritter. 1968. *Introduction to Chromatography*. New York: Reinhold Publishing Corp.

Lederer, E. and M. Lederer. 1957. *Chromatography. A Review of Principles and Applications*, 2nd edition. New York: American Elsevier Co.

Lehninger, A. L. 1982. *Principles of Biochemistry.* New York: Worth.

Macek, K. (ed.). 1972. *Pharmaceutical Applications of Thin Layer and Paper Chromatography*. Amsterdam: Elsevier.

Zweig, G. and J. Sherma (eds.). 1972. *Handbook of Chromatography*. Cleveland: Chemical Rubber.

EXERCISE 27

The precipitation and isoelectric point of proteins

Introduction

Through an understanding of the chemistry and reactivity of proteins, plant scientists have been able to isolate and purify these complex molecules of life. Some of the information obtained about the isolation techniques used in initial studies have shown that proteins can be extracted and precipitated out by use of various chemicals. You will observe the effects of precipitating agents of proteins in the following exercise. Proteins may be precipitated by heavy metals, yet we must remember that in the process, denaturation of the protein molecule results. Various alkaloidal reagents have also been used to isolate proteins. Another common method of isolating proteins, often with less harsh effects than heavy metals and alkaloidal reagents, is based on the phenomenon of "salting out," in which the protein is dehydrated and the charge is neutralized by a reagent such as ammonium sulfate with resulting precipitation.

At a certain pH, protein molecules may exhibit a net zero charge and will not migrate when subjected to electrophoresis. This form of the protein is due to the charge state of the surface amino acids that occur predominantly as zwitterions (NH_3^+ / COO^-). At the isoelectric point, proteins are somewhat easier to separate out because of the net zero charge at the surface.

Purpose

To demonstrate the effect of heavy metals and various reagents on the stability of proteins in solution and to determine the isoelectric point of legumin.

Materials

13 g pea seed
50 ml egg albumen (2%, w/v)
40 ml NaCl (0.4%, w/v)
60 ml NaCl (10%, w/v)
100 ml separate (1%, w/v) solutions of the following: $HgCl_2$, $AgNo_3$, $MgCl_2$, $CuSO_4$

NaOH (0.1N) in drop bottle
trichloroacetic acid (5%, v/v) in drop bottle
picric acid (saturated solution; explosive when dry)
tannic acid (10%, w/v) in drop bottle

phosphotungstic acid (5%, w/v) in
   drop bottle
8 ml ammonium sulfate (saturated
   solution)
5 g ammonium sulfate (solid)
1 ml acetic acid (concentrated)
Biuret reagent (see Appendix II)
50 ml sodium citrate (0.1M)
25 ml citric acid (1N)
25 ml citric acid (0.1N)
25 ml citric acid (0.1N)
cheesecloth (two layers)

4 centrifuge tubes (50 ml)
centrifuge
10 test tubes (maximum number
   per test)
water bath (boiling)
3 flasks (125 ml)
9 beakers (50 ml)
pH meter
9 pipets (10 ml graduated)
9 pipets (1 ml graduated)
several graduated cylinders (25 ml)

Procedure

A. Preparation of protein extract for precipitation experiments

Grind 6 g of dry pea seed (other seeds rich in protein may be used) in a cold mortar with 40 ml of ice-cold 0.4% NaCl solution for several minutes.

Strain the mixture through two layers of cheesecloth and centrifuge the filtrate at $1200 \times g$ for twenty minutes. Use the supernatant in the test outlined below. Also, for comparison purposes, perform the test on a 2% egg albumen solution.

B. Protein precipitation by heavy metals

Pipet 3 ml of test solution into each of four separate test tubes. To the first tube add drop by drop a 1% solution of $HgCl_2$. Note the result and then add an excess of 1% $HgCl_2$.

Repeat the above procedure for the three remaining test tubes, but in place of $HgCl_2$ substitute one of the following solutions: 1% $AgNo_3$, 1% $MgCl_2$, and 1% $CuSO_4$.

To 3 ml of test solution add 1% $CuSO_4$ drop by drop and, when no further precipitate forms, add two to three drops of 0.1N NaOH.

For the different solutions used, record the amount of precipitate on a relative basis (none, very slight, slight, and so on).

Observations:

C. Protein precipitation by alkaloidal reagents

Test the effects of 5% trichloroacetic acid, saturated picric acid, 10% tannic acid, and 5% phosphotungstic acid on separate 3 ml portions of pea extract and 3 ml portions of 2% egg albumen. Add the reagents drop by drop.

Observations:

D. Salting out

Add 2 ml of test solution to a test tube. Then add 2 ml of a saturated ammonium sulfate solution. Is any precipitate formed? Filter and perform the Biuret test on the filtrate (see Exercise 25).

To a second test tube containing 2 ml of test solution, add 2 ml of saturated ammonium sulfate. Then add sufficient solid ammonium sulfate to saturate the solution. Filter and perform the Biuret test on the filtrate.

To a 2 ml portion of test solution, add 2 ml of 10% NaCl solution. Then add two to three drops of concentrated acetic acid. Filter and test the filtrates with Biuret reagent.

Observations:

E. Isoelectric point of legumin

Protein extract

Grind 6 g of pea seeds in 40 ml of 10% (w/v) NaCl for several minutes. Transfer the mixture to a beaker and allow to stand with occasional stirring for fifteen minutes. Centrifuge at $1000 \times g$ for ten minutes and decant the supernatant into a flask. Place the flask into a boiling water bath and heat for fifteen minutes. Cool the flask under running tap water and recentrifuge the extract at $1000 \times g$ for ten minutes. Collect the supernatant and adjust the final volume to 50 ml with 10% NaCl solution. Use the extract as indicated under "Determination of isoelectric point of legumin" below.

Preparation of buffer solutions

Place 2 ml of 0.1M sodium citrate ($Na_3C_6H_5O_7 \cdot 2H_2O$) in each of nine 50 ml beakers (numbered 1 through 9). Then add 10 ml of water to each beaker. To adjust the solution in each beaker to the desired pH, add the appropriate amount and concentration of citric acid solution as follows:

1.　To beakers 1 and 2 add sufficient amounts of 0.1N citric acid to give a final pH of 7.0 and 6.5, respectively.

2.　To beakers 3, 4, 5, and 6 add sufficient amounts of 0.1N citric acid to give a final pH of 6.0, 5.5, 5.0, and 4.5, respectively.

3.　To beakers 7, 8, and 9 add sufficient 1N citric acid to give a final pH of 4.0, 3.5, and 3.0, respectively.

Adjust the final volume of each solution to 18 ml with distilled water and check the pH of each solution. Pour 9 ml of each buffer solution into separate and appropriately labeled test tubes so that the nine test tubes are serially arranged from high to low pH. The remaining buffer solutions may be used for a duplicate run if desired.

Determination of isoelectric point of legumin

To each of the nine solutions add exactly 1 ml of pea extract slowly. With frequent mixing, incubate the test tubes for twenty minutes at room temperature, and arbitrarily rate the amount of turbidity in each tube. The pH of the solution that exhibits the greatest relative turbidity represents the isoelectric point of the protein. If more than one solution exhibits relatively equal turbidity, interpolate from the pH values in question for an approximate isoelectric point value of the protein.

Observations:

Results/
Conclusions

Construct a suitable table illustrating your observations concerning the influence of heavy metals and other reagents on the stability of proteins in solution.

Construct a graph for the isoelectric point experiment in which the relative turbidities are plotted against the corresponding pH values.

In concluding statements, assess your results in terms of your knowledge of the physical and chemical properties of proteins. Particular emphasis should be directed to the nature of zwitterions, isoelectric point, protein precipitation by heavy metals and other reagents, salting out, and salting in.

Name three general biological roles attributed to proteins. Why are proteins excellent buffers? Would you expect a protein to be highly water soluble and physiologically active at its isoelectric point? Explain.

References

Anson, M. L. and J. T. Edsall (eds.). 1944–1968. *Advances in Protein Chemistry*, Vol. 1–23. New York: Academic Press.

Lehninger, A. L. 1982. *Principles of Biochemistry*. New York: Worth.

Metzler, E. D. 1977. *Biochemistry*. New York: Academic Press.

Neurath, H. (ed.). 1963–1966. *The Proteins: Composition, Structure, and Function*, 2nd edition, Vols. 1–5. New York: Academic Press.

Smith, H. (ed.). 1977. *The Molecular Biology of Plant Cells*. Berkeley: University of California Press.

Stahmann, M. A. 1963. Plant proteins. *Ann. Rev. Plant Physiol.* 14: 137–158.

Stryer, L. 1981. *Biochemistry*, 2nd edition. San Francisco: W. H. Freeman and Co.

EXERCISE 28

The pigments of the chloroplasts

Introduction

The compounds most important in the conversion of light energy to chemical energy are the pigments within the chloroplasts (or chromatophores) of plants. The pigments found in abundance in the chloroplasts of higher plants are the chlorophylls a and b and carotenoids. (Chlorophyll c, d, and e are found only in the algae and in combination with chlorophyll a).

The two major chlorophylls that will be extracted during the following exercise are chlorophylls a and b. Chlorophyll a usually appears blue-green; chlorophyll b, yellow-green. Further explanation of their structure has been considered within the introduction to Exercise 29.

Carotenoids are lipid compounds that are distributed widely in both animals and plants and range in color from yellow to purple. These pigments are present in different amounts in nearly all plants and in many microorganisms, including the red and green algae, photosynthetic bacteria, and fungi. Those carotenoids that consist exclusively of carbon and hydrogen are termed *carotenes*, while carotenoids that contain oxygen are called *xanthophylls*. Generally, the common names used to describe carotenes end in -ene and those of the xanthophylls end with -in. Carotenoids, like the chlorophylls, are located in the chloroplast or chromatophores and occur as water-insoluble protein complexes.

The major carotenoid found in plant tissues is the orange-yellow pigment β-carotene, which is generally accompanied by varying amounts of α-carotene. The chemical difference between α-carotene and β-carotene is that β-carotene consists of two β-ionone rings and α-carotene consists of one α- and one β-ionone ring.

The xanthophylls are more abundant in nature than are the carotenes. In growing leaves, for example, the concentration of xanthophylls may exceed that of the carotene by about two to one. Some of the major examples of xanthophylls occurring in plastids are lutein, violaxanthin, and neoxanthin.

The pigments of the chloroplast are readily soluble in acetone, which provides an excellent extraction medium for their isolation. Other organic solvents, such as carbon tetrachloride, petroleum ether, *n*-propanol, and so on are ideal for the chromatographic separation of chloroplast pigments. It is important to note, however, that any solvents used in the separation of plant products should be considered toxic and hazardous. Therefore, care should be taken in the use of organic solvents—they can also react with *your* chemicals. Take measures to avoid inhalation or contact with the skin.

Purpose

To extract and separate the chloroplast pigments by standard solvent washes and paper chromatography techniques.

Materials

1 g leaf tissue (spinach, blue grass, or other chlorophyll-containing tissue)

100 ml ethanol (95%)

20 ml acetone (85%, v/v)

10 ml acetone (absolute)

50 ml ethyl ether

50 ml Solvent A (carbon tetrachloride)

mixture of petroleum ether (1350 ml)

acetone (150 ml)

6.75 ml Solvent B (*n*-propanol)

$CaCO_3$ (solid)

Na_2SO_4 (anhydrous solid)

mortar and pestle

separatory funnel (250 ml)

ring stand and clamps

spectroscope (if available)

graduated cylinder (small)

2 bottles (small) with corks to fit

pint jar with screw top (chromatogram chamber)

chromatography paper (Whatman No. 1 or 3; small-squared to fit chamber)

support rod (glass tubing or meter stick)

glass tubing (drawn to a fine tip like a pen)

paper clips

ruler

eye dropper

Procedure

A. Pigment extraction

Note: Perform the extraction procedure under dim light conditions.

Weigh out approximately 1 g of green leaf tissue provided for the class, discarding large vein. Place the tissue with a very small amount of $CaCO_3$ (to neutralize cell acids and prevent the removal of the magnesium from the chlorophyll nucleus) in a clean mortar and grind to a fine pulp. Add enough 85% acetone to thin the pulp (5 to 6 ml will suffice) and continue to grind. Using an eye dropper with a fine tip, transfer the clear supernatant green liquid to 10 ml of ethyl ether contained in a 250 ml separatory funnel.

Repeat the extraction procedure on the tissue a second time and add the resulting liquid to the separatory funnel containing the first extract and the ethyl ether. After the second extraction, add about equal parts of absolute acetone and ethyl ether to the tissue residue. Grind the tissue and transfer the fluid to the separatory funnel as before. Then make a last extraction with the ethyl ether. The ether facilitates the extraction of the yellow pigments. After this extraction, be sure to transfer the clear green liquid to the separatory funnel.

The green solution in the separatory funnel contains the plastid pigments together with small amounts of other compounds in a mixture of acetone materials that are water soluble. Now add 100 ml of distilled water to the pigment solution in the separatory funnel. Add the water slowly, by pouring it down the funnel to avoid the formation of an emulsion. Rotate the funnel—*do not shake*—for a few minutes to speed the transfer of the acetone and other substances into the lower water layer.

Fasten the separatory funnel in an upright position. When two liquid layers are sharply defined, run off the lower layer and discard it. Repeat the washing of the ether solution with water three more times to remove all of the acetone. Since ether is soluble in water to some extent, it may be necessary to add 5 to 10 ml of ether to replace that removed in the water wash. However, do not add more ether as long as there are two distinct liquid layers in the separatory funnel since it is desirable that the volume of the ether layer does not exceed 15 ml at the beginning of each washing.

After the water washes, collect the ether solution of the chloroplast pigments into a small graduated cylinder. If necessary, add ether to increase the total volume of the solution to 10 ml. If the volume is more than 10 ml, transfer the solution to a beaker and allow enough ether to evaporate so that the solution can be fixed at this volume. Observe the ether solution with transmitted and reflected light.

Pour the pigment solution into a small bottle containing 2 g of *anhydrous* Na_2SO_4. Cork the bottle and swirl the contents so that the salt is suspended in the liquid. Continue swirling for several minutes. The anhydrous salt serves as a dehydrating agent to remove the residual water in the ether. The extract is ready for further separation techniques after the salt has settled.

B.   Separation of plastid pigments by paper chromatography

Prepare the chromatographic chambers by pouring carbon tetrachloride (Solvent A) to a depth of one-half inch in a pint jar. Carbon tetrachloride is *very toxic*, and all procedures with it should be performed in a hood. Larger chromatogram jars may be used if desired. Add anhydrous sodium sulfate and close the jar tightly with a screw top.

A second solvent system (B) that may be used is a mixture of petroleum ether, acetone, and *n*-propanol (*Careful! Toxic! Explosive! Use hood and rubber gloves*). In using this system, omit the anhydrous sodium sulfate in the chamber and use Whatman No. 3 paper instead of Whatman No. 1.

While the internal atmosphere of the jar is equilibrating, cut Whatman No. 1 (or No. 3) chromatography paper into square sheets that will fit easily into the jar when rolled into a cylinder. Draw a pencil line approximately 1 in. from the bottom of each sheet and store over a desiccant if not used immediately.

Arrange the sheet to overlap a support such as a glass tube or meter stick in such a way that the pencil line is raised above the surface of the table.

Using a glass tube drawn to a fine tip like a pen, draw a line with the pigment extract for about $1\frac{1}{2}$ in. along the pencil mark on each paper. Allow to dry and repeat the application several times, or until the mark is deep green. Keep the

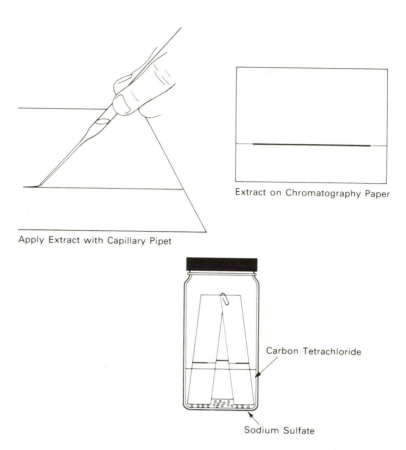

Apply Extract with Capillary Pipet

Extract on Chromatography Paper

Carbon Tetrachloride

Sodium Sulfate

FIGURE 28–1   Preparation of chromatograms for pigment separation.

pigment line as narrow as possible and allow the pigment to dry between each application.

When the pigments are dry, roll the paper into a loose cylinder with the marked end down, secure with a paper clip, and stand it in the jar containing carbon tetrachloride (or Solvent B) and an excess of anhydrous sodium sulfate on the bottom (see Figure 28–1).

The solvent will move up the paper. Remove the chromatogram when the solvent nears the top of the paper (in about one hour) and allow to dry.

Examine the paper when it is fresh since the colors will fade, especially in the light. The pigments should appear as colored bands in the following sequence from top to bottom: carotenes (orange-yellow at the solvent front), xanthophylls (one or more yellow bands), chlorophyll a (blue-green), chlorophyll b (yellow-green). The instructor may suggest other chromatographic solvents to try. In that case, repeat the same procedures but substitute the desired solvent for the carbon tetrachloride. Remember, with different chromatographic solvents the patterns of pigment separation may be different.

Results/
Conclusions

Calculate the Rf values of the separated pigment bands on the chromatograms. In concluding statements, consider the following:

1. The phenomena involved in the separation of the pigments on paper (principles of chromatography).

2. The structural formulas for chlorophyll a, chlorophyll b, a xanthophyll, and α-carotene, and where these pigments are located in plant cells.

3. The location of anthocyanin pigments in the cell.

4. Those pigments that develop in the organs of plants grown in darkness and the color changes plastids may undergo.

Write the saponification reaction and explain the change in solubility when the chlorophylls are treated with strong alkali, for example, KOH. Define fluorescence and phosphorescence.

Acknowledgment

This exercise was designed for class use by C. W. Hagen, Jr., Indiana University, Bloomington.

References

Ellis, J. 1981. Chloroplast proteins: Synthesis, transport, and assembly. *Ann. Rev. Plant Physiol.* 32: 111–137.

Goodwin, T. 1960. Chemistry, biogenesis, and physiology of the carotenoids. In Richland, W. (ed.), *Encyclopedia of Plant Physiology*, Vol. 5. Berlin: Springer–Verlag, 394–443.

Goodwin, T. 1966. *Biochemistry of Chloroplasts*, Vols. 1 and 2. New York: Academic Press.

Heber, U. and H. W. Heldt. 1981. The chloroplast envelope: Structure, function and role in leaf metabolism. *Ann. Rev. Plant Physiol.* 32: 139–168.

Jensen, A. and O. Aasmundrud. 1963. Paper chromatographic characterization of chlorophylls. *Acta Chem. Scand.* 17: 907–912.

Kirk, J. T. O. and R. A. E. Tilney-Bassett. 1978. *The Plastids: Their Chemistry, Structure, Growth and Inheritance*. New York: Elsevier North-Holland.

Possingham, J. V. 1980. Plastid replication and development in the life cycle of higher plants. *Ann. Rev. Plant Physiol.* 31: 113–129.

EXERCISE 29

Chlorophyll absorption spectrum and qualitative determinations

The chlorophyll pigments, which play an important role in the photosynthetic process, consist of a cyclic tetrapyrrolic structure (porphyrin) with an isocyclic ring containing a magnesium atom at its center. In addition, a phytol chain of the chlorophyll extends from one of the pyrrole rings. The phytol tail is a long hydro-carbon chain that contains one double bond and is esterified with the carobxyl group of one of the pyrrole rings of the porphyrin.

Although there are at least nine types of chlorophyll (a, b, c, d, e, bacterio-chlorophyll a and b, and two chlorobium chlorophylls), chlorophyll a and b are the most abundant and are found in all autotrophic organisms except the pigment-containing bacteria. Chlorophyll b is also absent in the blue-green, brown, and red algae. The major chemical difference between chlorophyll a and b is that in chlorophyll a there is a methyl group (CH_3) linked to the third carbon of the porphyrin ring, and in chlorophyll b an aldehyde group (HC = O) is linked to the third carbon.

Chlorophyll a and b each exhibit different absorption spectra—that is, mea-sures of the amount of light absorbed by a given substance exposed to different wavelengths. Since an absorption spectrum depends on the unique absorption characteristics of a compound, scientists use it to identify substances with a great deal of accuracy. The following procedures will illustrate methods for determining the total amount of chlorophyll a and b and the absorption spectrum of chloro-phyll extracted from beans.

Both chlorophyll a and b show maximum light absorption in the blue-violet region and orange-red region of the visible spectrum. They also show minimum absorption of the green and yellow wavelengths. These absorption properties will be illustrated and utilized to make quantitative as well as qualitative determina-tions of chlorophyll extracted from spinach leaves.

Purpose
To study a technique used to determine the amount of total chlorophyll, chlorophyll a, and chlorophyll b present in leaves; to determine the absorption spectrum of chlorophyll extracted from leaves.

Materials

1 g spinach leaves (or other fresh leaf material)	Buchner funnel, containing a pad of Whatman No. 1 filter paper
105 ml acetone (80%, v/v)	side-arm suction flask
mortar and pestle (homogenizer may be used)	cuvettes (10 mm)
	spectrophotometer

Procedure

A. **Chlorophyll extraction**

Place 1 g (fresh weight) of small cut pieces of spinach leaves (or other fresh green leaves) into a clean mortar. Add 40 ml of 80% (v/v) acetone and grind the tissue to a fine pulp (for about three minutes).

Carefully transfer the resulting green liquid to a Buchner funnel containing a pad of Whatman No. 1 filter paper. While filtering the extract with suction, repeat the grinding of the pulp with a fresh 30 ml aliquot of 80% acetone. After three to four minutes, filter the second extract as before into the flask containing the first extract.

After the second extraction the tissue should be devoid of chlorophyll. If not, repeat the grinding with a fresh 20 ml aliquot of 80% acetone. Then filter the slurry into the flask containing the other filtrates. With 10 ml of 80% acetone rinse the mortar and sides of the funnel to ensure that all the chlorophyll is collected. For convenience of calculating the amount of chlorophyll present, adjust the final volume of the filtrate to 100 ml by adding sufficient 80% acetone.

If experimental plant material is to be extracted and less than a 1 g sample is available (and for convenience later), adjust the amounts of 80% acetone used in the extraction and the final volume accordingly so that the final extract has a volume based on 10 mg plant material extracted per 1 ml of acetone.

B. **Chlorophyll determinations**

Read and record the optical density of the chlorophyll extract in a 10 mm cell with a spectrophotometer set at 645, 663, and 652 nm. Be sure to read against an 80% acetone solvent blank. Then calculate the amount of chlorophyll present in the extract on the basis of milligrams of chlorophyll per gram of leaf tissue extracted according to the following equations:

$$\text{mg chlorophyll a/g tissue} = [12.7(D_{663}) - 2.69(D_{645})] \times \frac{V}{1000 \times W}$$

$$\text{mg chlorophyll b/g tissue} = [22.9(D_{645}) - 4.68(D_{663})] \times \frac{V}{1000 \times W}$$

$$\text{mg total chlorophyll/g tissue} = [20.2(D_{645}) + 8.02(D_{663})] \times \frac{V}{1000 \times W}$$

$$\text{mg total chlorophyll/g tissue} = \frac{D_{652} \times 1000}{34.5} \times \frac{V}{1000 \times W}$$

where:

D = optical density reading of the chlorophyll extract at the specific indicated wavelength

V = final volume of the 80% acetone-chlorophyll extract

W = fresh weight in grams of the tissue extracted

Calculations and chlorophyll determinations:

C. Absorption spectrum of chlorophyll in 80% acetone

Dilute a portion of the extract used above with sufficient 80% acetone to obtain a transmittance reading of about 30 to 35%. Be sure to use an additional cuvette filled with 80% acetone for a reagent blank.

Take and record the optical density readings of the diluted chlorophyll extract at 20 nm intervals over a range of 350 to 700 nm. At points of maximum absorption, take additional readings at 5 nm intervals. In addition, observations of visible light absorption by the chlorophyll extract may also be performed with a hand spectroscope (if available).

Optical density readings:

Results/
Conclusions

Present the calculated amounts of chlorophyll in milligrams per gram of leaf material extracted. Also, include a graph illustrating the absorption spectrum of chlorophyll in which the optical density readings are plotted on the ordinate against the wavelengths on the abscissa. In concluding remarks, consider the following:

1.   Why is 80% acetone used as the extraction medium and as the reagent blank in the spectrophotometric determinations?

2.   With respect to the equations used in calculating the amount of chlorophyll, what is the importance of the absorption coefficients used, and why were the optical density readings taken at 645, 663, and 652 nm used?

3.   Is the method for chlorophyll determination presented in this exercise subject to error? Explain.

4.   Of what importance is an absorption spectrum as it relates to natural product identification?

5.   What is the difference between an absorption spectrum and action spectrum, and how are they used to obtain insight into light-mediated responses in plants?

Acknowledgment

The absorption coefficients were taken from the work of G. MacKinney and the equations for chlorophyll determination from the work of D. I. Arnon (see the references for this exercise).

References

Arnon, D. I. 1949. Copper enzymes in isolated chloroplasts. Polyphenoloxidase in *Beta vulgaris. Plant Physiol.* 24: 1–15.

Bray, J. R. 1960. The chlorophyll content of some native and managed plant communities in central Minnesota. *Can. J. Bot.* 38: 313–333.

Bruinsma, J. 1961. A comment on the spectrophotometric determination of chlorophyll. *Biochem. Biophys. Acta* 52: 576–578.

Castelfranco, P. A. and S. I. Beale. 1983. Chlorophyll biosynthesis: Recent advances and areas of current interest. *Ann. Rev. Plant Physiol.* 34: 241–278.

Gregory, R. P. F. 1977. *Biochemistry of Photosynthesis*, 2nd edition. New York: John Wiley and Sons.

Haliwell, B. 1981. *Chloroplast Metabolism*. Oxford: Oxford University Press.

Hipkins, M. F. 1984. Photosynthesis. In Wilkins, M. B. (ed.), *Advanced Plant Physiology*. London: Pitman Publishing, Limited, pp, 219–248.

MacKinney, G. 1938. Some absorption spectra of leaf extracts. *Plant Physiol.* 13: 128–140.

MacKinney, G. 1941. Absorption of light by chlorophyll solutions. *Biol. Chem.* 140: 315–322.

Zscheile, F. P. and C. L. Comar. 1941. Influence of preparative procedure on the purity of chlorophyll components as shown by absorption spectra. *Bot. Gaz.* 102: 463–481.

EXERCISE 30

The separation and identification of lycopene

Introduction

The natural carotenoids are derivatives of lycopene, a red pigment found in abundance in the plastids of tomatoes and in many other plants. Lycopene is a highly unsaturated straight-chain hydrocarbon composed of two identical units joined by a double bond between carbon atoms 15 and 15'. The empirical formula is $C_{40}H_{51}$. Like most carotenoids, it is composed of eight isopene-like residues with each half of the molecule derived from four isopene units (isopene has the formula $CH_2 = C(CH_3)–CH = CH_2$). The molecular structure of lycopene is as follows:

Lycopene

For further information on carotenoids, see the introduction to Exercise 28 and the textbooks cited in the references list. While performing the exercise, be sure to consider any reagents used in the extraction and preparation of plant products as potentially hazardous. Inhalation of vapors and contact with the skin should be assiduously avoided.

Purpose

To extract the principal pigments of the pulp of ripe tomatoes—carotenoids, mainly lycopene; to separate them by column chromatography and identify the lycopene by determining its absorption spectrum.

Materials

tomatoes (ripe)	Buchner funnel with side-arm suction flask
150 ml acetone (toxic)	separatory funnel (500 ml)
300 ml benzene (toxic)	evaporating device (see Figure 30–1A)
5 ml ethanol (absolute)	pipet (Pasteur, rubber bulbs or disposable)
50 ml ethanol (95%)	column chromatography apparatus (see Figure 30–1B)
Na_2SO_4 (anhydrous solid)	cotton
$CaCO_3$ (anhydrous solid, pre-baked)	tamper (for packing column)
Al_2O_3 (alumina, pre-baked)	3 evaporating dishes
blender	cuvettes
beaker (500 ml)	spectrophotometer
Whatman No. 1 filter paper disks (to fit Buchner funnel and column)	

Procedure

A. Lycopene extraction

Homogenize one-half of a ripe tomato (remove rotten spots and the green stem) in a blender. Add 100 ml of acetone containing a pinch of $CaCO_3$ and 100 ml of benzene and blend the contents again for several minutes. Pour the homogenate into a 500 ml beaker and allow to settle.

Filter the contents using suction through a pad of Whatman No. 1 filter paper by means of a Buchner funnel into a 500 ml filter flask. The process will go faster if the pulp is not poured into the filter until all the liquid is sucked through.

Press and wash the pulp several times with a few ml of acetone until it is dry and relatively colorless. Do *not* use over 50 ml of acetone for the washing. Discard the pulp and save the filtrate.

The filtrate will probably be in two layers. Add 200 ml of water to the filtrate and swirl gently to mix the water with the lower layer. Pour the whole mixture into a 500 ml separatory funnel and discard the bottom water-acetone layer, which is light yellow.

Keep and wash the benzene upper layer (orange in color) with 100 ml portions of water until acetone can no longer be detected in the separated water layer. This usually takes two or three washings depending on how much acetone was used during filtration. During the washings the solutions should be swirled, *not shaken*, or an emulsion will form that takes a long time to settle.

Pour the benzene solution into a 150 ml Erlenmeyer flask equipped with a stopper and tubing as indicated in Figure 30–1A. The benzene can be evaporated quickly (to about 1 cm deep) by heating the solution in a water bath at 30 to 35°C while drawing air over it with a water pump. Do not concentrate the solution too

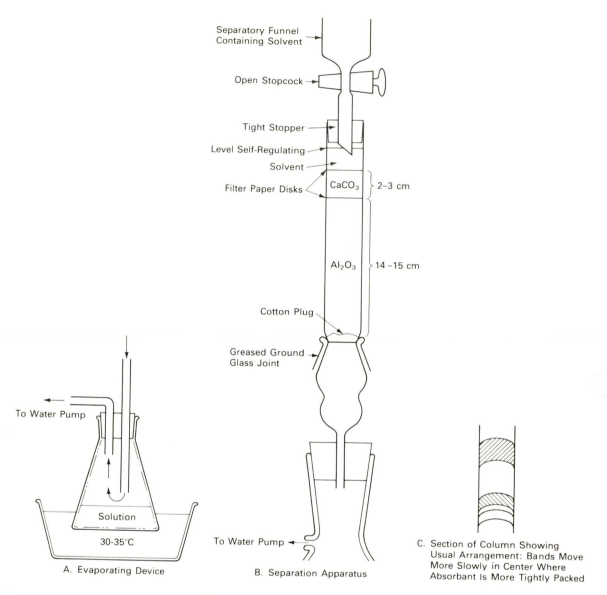

Separatory Funnel
Containing Solvent

Open Stopcock

Tight Stopper

Level Self-Regulating

Solvent

Filter Paper Disks

CaCO₃ } 2-3 cm

Al₂O₃ } 14-15 cm

Cotton Plug

Greased Ground
Glass Joint

To Water Pump

Solution

30-35°C

A. Evaporating Device

To Water Pump

B. Separation Apparatus

C. Section of Column Showing
Usual Arrangement: Bands Move
More Slowly in Center Where
Absorbant Is More Tightly Packed

FIGURE 30-1   Apparatus for lycopene extraction.

much or crystals that are difficult to redissolve will separate. On the other hand, if the solution is too dilute, later separation will be poor.

When the solution is concentrated, a small amount of anhydrous sodium sulfate is added to dry it. Store in the refrigerator to prevent oxidation by heat and light.

B.   Column chromatography

The setup for chromatography is shown in Figure 30–1B. Insert a suitable column with a cotton plug in the bottom into a suction flask and turn the suction on. Pour pre-baked alumina into the tube in small portions and press down firmly with rotary motion of the tamper. It is very important to have even packing in the column; otherwise, the bands will not be flat and will be difficult to separate. Continue to add and pack the alumina until it is 14 to 15 cm deep in the column. On the top of the alumina place a small disk of filter paper (same diameter as the column) to keep a sharp boundary. Place 2 to 3 cm of pre-baked calcium carbonate on top of the filter paper disk. Do not pack the calcium carbonate too tightly or else it may prevent the passage of solvent. Place another filter paper disk on top of the whole column.

Add benzene to the column and apply suction to pack the column tighter and to remove pockets of air. When the entire column is wetted with benzene, add the concentrated pigment solution through a pipet. Be careful to get the whole upper surface covered at the same time and to keep it covered. Do not run in too much solution at once since evaporation of the solvent will deposit pigment on the walls of the tube and the pure benzene will later be contaminated, giving poor separation. Also, the pigment should not be spilled down the walls of the tube, but delivered right on top of the paper disk in the center. Add enough solution to give a band on the alumina about 1 cm thick. Then pour in anhydrous benzene and attach a separatory funnel containing benzene on top of the column as indicated in Figure 30–1B.

Run anhydrous benzene through the column *slowly* (several drops at a time). Make sure any pigment deposited on the walls of the tube by evaporation is dissolved by the first portion of the pure benzene. Keep the top of the column covered with benzene at all times and never stop suction. The funnel will feed automatically if it is fitted to the column with a tight stopper.

The pigments should separate into five main bands in the following order:

1.   Yellow band (which contains α and β carotenes and will wash right into the flask).

2.   Orange band (which separates very slowly below band 3).

3.   Thick red band (which will be contaminated on its lower side with band 2).

4.   Red band (near the top).

5.   Brown band (on the very top of the alumina).

At first there may be thin yellow bands on the calcium carbonate. They will wash down and merge with band 5.

Continue the development until bands 2 and 3 appear to separate fairly well. Separation requires about 180 ml benzene and should not last over an hour. If progress is too slow, add two drops of absolute ethanol/100 ml dry benzene.

After bands 2 and 3 have separated fairly well, allow the column to suck dry for half a minute and quickly detach the column. Punch out the cylinder of alumina onto paper toweling and carefully dig out the bands separately with a spatula. Band 3 is lycopene, and bands 2 and 4 are isomers of lycopene. Save only the top half of band 3 since the bottom half is contaminated with the component of band 2. It is important to remember that a cross section of the column usually looks like Figure 30–1C. Bands 2 and 4 give good absorption curves and may be saved for comparison to lycopene.

Place each of the alumina bands in a small beaker or evaporating dish and cover with ethanol. Stir and filter the liquid through a piece of filter paper, collecting the eluted pigment in a small flask or bottle. If the pigment was not all eluted, add more ethanol and repeat, but avoid using too much ethanol and getting too dilute a solution. Add 20 to 30 ml of benzene to the ethanol eluates, place in a separatory funnel, and remove the alcohol by repeated washing with water. Dry the benzene solution with anhydrous sodium sulfate. If it has to be stored, leave it in the refrigerator.

C. Absorption banding of lycopene

Before using the spectrophotometer, acquaint yourself with the instrument and the operating instructions. Then fill one of the cuvettes with pure benzene (toxic) to be used as the solvent blank for standardizing the instrument, and place the pigment solution in another cuvette. The pigment solution may be too concentrated. Test the absorption at 470 nm. It should be between 30 and 50% transmission, but certainly not less than 10%. If the solution is too concentrated, dilute it by adding benzene until it is in the proper range.

Measure the absorption from 420 to 520 nm. The best way is to start at 420 nm and take readings every 4 nm until the range is covered. Find where the density maxima (transmission minima) are and take readings every nm near the maximum until you find the peak absorption. There are three optical density maxima for lycopene: one near 455 nm, one near 470 nm, and one just over 500 nm.

Results/
Conclusions

Plot the absorption spectrum by recording the optical density for each reading and plot the values on the ordinates against the respective wavelengths on the abscissa.

Compare your lycopene curve with published ones (see Karrer and Jucker, *Carotenoids*). Solutions in ethanol, benzene, and hexane all give the same shape curves with nearly the same maxima.

Acknowledgment

This exercise was adapted for class use by C. W. Hagen, Jr., Indiana University, Bloomington.

References

Goodwin, T. 1960. Chemistry, biogenesis and physiology of the carotenoids. In Richland, W. (ed.), *Encyclopedia of Plant Physiology*, Vol. 5, Part 1. Berlin: Springer-Verlag, 394–443.

Haliwell, B. 1981. *Chloroplast Metabolism*. Oxford: Oxford University Press.

Karrer, P. and E. Jucker. 1950. *Carotenids*. New York: Elsevier Publishing Co.

Kirk, J. T. O. and R. A. E. Tilney-Bassett. 1978. *The Plastids: Their Chemistry, Structure, Growth and Inheritance*. New York: Elsevier North-Holland.

Malkin, R. 1982. Photosystem 1. *Ann. Rev. Plant Physiol.* 33: 455–479.

Zscheile, F. P., J. W. White, B. W. Beadle, and J. R. Roach. 1942. The preparation and absorption spectra of five pure carotenoid pigments. *Plant Physiol.* 17: 331–346.

EXERCISE 31

Anthocyanins from flowers of different stages of maturation

Introduction

The anthocyanins are widespread in plants and are often responsible for the violet, blue, or red colors characteristics of many fruits, flowers, and leaves (particularly in leaves during the autumn months). These pigments, which are primarily found in the large central vacuoles of parenchyma cells, are highly soluble in water because they are sugar derivatives (glycosides) of substances termed anthocyanidins.

The general anthocyanidin structure is as follows:

The anthocyanins are produced by sugar attachment to the hydroxyl position indicated (– OH). The different colors of the anthocyanins are due to the extent of methylation and hydroxylation of the B-ring.

Anthocyanidins and their derivative, the anthocyanins, can be separated by thin-layer and paper chromatography (for a discussion of paper chromatography, see the introduction to Exercise 21). Various phenolics anthocyanins, and anthocyanidins have been studied in the past with respect to their relative frequency of occurrence in different species of plants. In fact, earlier studies established the basis for "tracking" molecules in an approach currently known as chemotaxonomy.

The idea behind chemotaxonomy is that closely related plants contain similar and/or identical compounds. In addition, plant scientists have followed the development sequences involved in anthocyanin formation by extracting the pigments at different times during flower development. In this way much information has been obtained concerning the genetic regulation of the biosynthesis and interconversions of these pigments.

The following experiments will illustrate some of the early techniques involved in studying pigment formation during different stages of flower formation.

Purpose

To extract and separate anthocyanins by means of paper and/or thin-layer chromatography; to observe the changes that take place with respect to the production of various anthocyanins and anthocyanidin derivatives in flower petals during maturation.

Materials

Impatiens balsamina L. plants: four or five flower buds from each of three different stages of maturation (petunia, phlox, pansy, or four o'clock may be used)

10 ml HCl (1%, v/v) in 95% ethanol

300 ml Solvent I (*t*-butanol–acetic acid–water, 3:1:1)

300 ml Solvent II (10%, v/v; acetic acid in water)

300 ml Solvent III, (*n*-butanol–acetic acid–water, 5:1:4)

NH$_4$OH (concentrated)

AlCl$_3$ spray reagent (see Appendix II)

atomizer and bottle (or other spraying device)

scalpel or razor blades

mortar and pestle

glass wool, glass funnel for filtering

evaporating device (see Figure 30–1A)

micropipets (or glass tubing drawn to a fine point)

paper chromatography equipment (paper, stapler and staples, chromatocab, chromatography jar with lid, ruler, grease)

ultraviolet light source

thin-layer chromatography equipment (glass plates, metal holders for plates, spreader, square chromatography jar and lid, cellulose, silica gel H)

hood (for drying and spraying chromatograms)

Procedure

A. Extraction of anthocyanin derivatives from flower petals of different ages

From *Impatiens balsamina* L. plants growing in the greenhouse, pick four to five buds of three different stages of maturation:

Stage 1: closed bud when there is only a slight trace of color in the petals

Stage 2: closed buds that look "full" and where there is color showing through the sepals

Stage 3: deeply colored, fresh, fully opened flowers

Prepare three separate extracts by homogenizing the buds of each stage (1 and 2) and the petals of stage 3 with a small volume (2 to 3 ml) of 1% HCl in 95% ethanol.

Filter the liquid of each homogenate through glass wool and concentrate to near dryness or until a precipitate begins to form. The filtrates may be used immediately or stored in a refrigerator prior to their separation by paper or thin-layer chromatography.

The basic techniques outlined below may be highly modified in consideration of the facilities available and at the discretion of the laboratory instructor. For example, extracts from other plant tissues containing anthocyanin may be used as well as different chromatographic solvents. If commercial battery jars or chromatographic chambers are not available, they can easily be made from discarded quart-size pickle jars and even smaller sized screw-top jars if so desired. In addition, if two-dimensional paper chromatography is not feasible, then these extracts may be applied to one paper and developed in 10% acetic acid by ascending chromatography.

B.   Paper chromatography (two-dimensional)

Place a large sheet of Whatman No. 1 chromatography paper on a clean desk top. Measure 4 in. up from the bottom and 2 in. in from the left-hand margin, and make a small "X" with a pencil. With a drawn glass pipet, apply a drop of extract from stage 1 on the pencil mark. The spot should be no larger than 5 to 6 mm in diameter and should be allowed to dry before the next application. Continue to spot on the "X" mark for ten to twenty applications or until the spot may be considered "heavy." Be sure to allow at least a minute between applications so that the liquid will have a chance to dry.

Repeat the same application procedure with a fresh sheet of chromatography paper for each of the two remaining extracts (stages 2 and 3).

After the extracts have been applied, place the papers in a chromatocab with the spot at the top and develop by descending chromatography using a solvent system consisting of t-butanol–acetic acid–water (3:1:1). At the end of the run (approximately twelve hours), hang the papers to air-dry with the point of origin end down. Mark the solvent front at this time.

After the papers are dried (about one hour), observe the separated components under daylight and short-wave ultraviolet light (*Caution! Do not look directly at the lamp!*) and place a pencil dot in the center of each. Determine and record the Rf for each component and note their color in visible light or if they fluoresce under ultraviolet light.

For a second dimensional run, orient the papers so that the origin of the first run ("X" pencil mark) is on the extreme right-hand side and the solvent front is to the left and perpendicular to the line of separated components. Then roll each paper into a cylinder (with the spots at the bottom and on the external side of the cylinder) and staple the edges. Stand the cylinders upright in battery jars containing sufficient solvent (10% acetic acid) to wet the cylinder just below the previously separated components.

The time for development in 10% acetic acid is between one and one-half and two hours. When the solvent front approaches within 1 to 2 in. from the top edge of the cylinder, remove the papers from the jar and hang them to dry.

Observe the dried papers under visible and ultraviolet light and mark the colored or fluorescent components in the manner previously described. Individual spots may become clearly defined by holding the paper over an open bottle or flask of concentrated ammonium hydroxide. If desired, the papers may also be sprayed with 2% AlCl$_3$, and after drying observed again under visible and short-wave ultraviolet light.

C. Thin-layer chromatography

Preparation of plates

Prepare a slurry consisting of a mixed layer of cellulose-silica gel H (10 g cellulose blended with 4 g silica gel H and 80 ml of distilled water; sufficient for five 20 × 20 cm plates). When the slurry is mixed and is homogeneous, prepare the plates with the spreader provided. The spreader gate should be adjusted so that the thickness of the cellulose-gel layer is approximately 250 microns. After spreading, place the plates on a smooth surface, tap the ends gently several times, and when the plates have set, oven dry them at 40°C overnight.

As an alternative method, plates may be prepared with a slurry consisting of approximately 1 part adsorbosil-2 and 1 part water. (This gel layer does not give the resolution obtained with the previously described medium, however.) If silica gel (adsorbosil-2) is used, the plates should be activated for thirty to thirty-five minutes in an oven set at 110 to 120°C. The plates can be used after they are removed from the oven, placed in a dessicator box, and allowed to cool to room temperature.

Application and development

Carefully spot 3 to 6 μl of extract from stage 1 about an inch from the bottom and an inch from the left hand margin of a plate. Further to the right and on the same line of origin, apply the same quantity of extract from stage 2, and then further to the right the extract from stage 3. The stages should be represented on one plate as three distinct spots having approximately the same origin.

Develop the plates in Solvent III, *n*-butanol–acetic acid–water (5:1:4). Pour sufficient solvent into a square chromatography jar to a level of $^1/_2$ in. from the bottom. Mount the spotted plate vertically in the chromatography jar using the metal holders provided so that the spots are just above the solvent level. Cover the jar with a suitably greased lid and allow one to two hours for development. When the solvent front reaches a point 1 in. below the top of the plate, remove it from the jar and allow to dry.

The original sample mixture of each stage should resolve into several distinct spots. Observe and mark the separated components under visible and short-wave ultraviolet light. (*Caution! Protect your eyes!*) If desired, a permanent record of the separation may be obtained by placing a clean plate over the chromatograms.

Then make an overlay and trace the various components separated. Also, calculate the Rf values and note whether there is any correlation between the number of anthocyanin derivatives separated and the stage of flower development.

If recovery of the pigments is desired, they can be removed from the plate by carefully scraping the adsorbent into a beaker and eluting with 1% HCl in methanol, followed by filtration through glass wool or centrifugation and collection of the pigment-containing eluate. If desired, the adsorption spectrum for separated pigments may also be determined.

Results/ Conclusions

Design and present a suitable illustration for the above separation of anthocyanin derivatives from different stages of buds. If a table is used, be sure to include the method of chromatography used, the solvent system, the number of components separated from each extract and their Rf values, and whether the components are brightly colored in visible light and/or fluorescent under short-wave ultraviolet light. An overlay of the chromatograms may be included for further illustration of results.

Although no positive identification of the various separated components was attempted, the number and pattern of red and ultraviolet fluorescing spots obtained from the three stages should provide some appreciation for the pigment changes that take place during flower development in *Impatiens*. Your concluding statements should consider these changes in pigment formation as well as the basic principles concerning the separation of components by paper and thin-layer chromatography (if not considered in past experiments).

Acknowledgment

This exercise was adapted from the work of R. L. Mansell and C. W. Hagen, Jr.

References

Drumm, H. and H. Mohr. 1974. The dose response curve in phytochrome-mediated anthocyanin synthesis in the mustard seedling. *Photochem. Photobiol.* 20: 151–157.

Goodwin, T. W. (ed.). 1965. *Chemistry and Biochemistry of Plant Pigments.* New York: Academic Press.

Hagen, C. W., Jr. 1959. Influence of genes controlling flower color on relative quantities of anthocyanins and flavonols in petals of *Impatiens balsamina. Genetics* 44: 787–793.

Hagen, C. W., Jr. 1966a. The differentiation of pigmentation in flower parts. I. The flavonoid pigments of *Impatiens balsamina*, genotype 11HHPP and their distribution within the plant. *Amer. J. Bot.* 53: 46–54.

Hagen, C. W., Jr. 1966b. The differentiation of pigmentation in flower parts. II. Changes in pigments during development of buds in *Impatiens balsamina*, genotype 11HHPP. *Amer. J. Bot.* 53: 54–60.

Harborne, J. R. 1958. Spectral methods of characterizing anthocyanins. *Biochem. J.* 70: 22–28.

Mansell, R. L. and C. W. Hagen, Jr. 1966. The differentiation of pigmentation in flower parts. III. Metabolism of some exogenous anthocyanins by detached petals of *Impatiens balsamina. Amer. J. Bot.* 53: 875–882.

Miles, C. D. and C. W. Hagen, Jr. 1968. The differentiation of pigmentation in flower parts. IV. Flavanoid elaborating enzymes from petals of *Impatiens balsamina. Plant Physiol.* 43: 1347–1354.

Cell wall substances

Introduction

Cell walls exhibit important functions that are a part of the dynamic interactions between the external environment and the protoplast. The rigidity of the cell wall as well as the pressure of water in the vacuoles of plant cells are important in maintaining cellular integrity and mechanical support. Other functions of cell walls include involvement in absorption, transport of water and minerals, secretions, certain enzymatic activities, and growth. It is also suggested that cell wall components play an important role in disease resistance of plants.

Components of the cell wall are produced by the protoplast and deposited adjacent to the external surface of the plasmalemma. The major component of the cell wall is cellulose, a polysaccharide consisting of β-D-glucose. Pectic substances, hemicelluloses, lignin, cutin, suberin, and proteins, including enzymes, are the other major components of cell walls. Further information on the chemistry of cell wall components and its structure may be found in the texts cited in the references at the end of this exercise.

Purpose

To study the distribution of various cell wall components by staining procedures and microscopic observations.

Materials

tomato, sunflower, or maple (young stems)

cotton

nasturtium seed (separate portions of dry seed, seed soaked in water for several hours, and seed germinating for 48 hours)

Aloe or *Clivia* leaves

geranium stem

I_2KI reagent (see Appendix II)

H_2SO_4 (60%, v/v)

cuprammonia reagent (see Appendix II)

ruthenium red solution (see Appendix II)

alcoholic phloroglucinol solution (1 g of phloroglucinol in 100 ml of 95% ethyl alcohol)

Sudan III reagent (see Appendix II)

HCl (12%, v/v)

HCl (concentrated)

microscope slides and coverslips

microscope

filter paper

(All reagents only in quantities for separate dropping bottles)

Procedure   **A.   Detection of cellulose**

Cross sections of young stems

Mount a cross section of a young stem of tomato, sunflower, or maple on a slide in a drop of I_2KI reagent. After a minute, remove the I_2KI solution with a strip of filter paper and add a drop of 60% (v/v) sulfuric acid to the section. Add a coverslip and observe at once under a microscope. Cellulose structures should swell and stain blue. Draw the section observed and indicate those tissues composed largely of cellulose.

Observations:

Structural organization of cellulose in plant cell walls

Mount a few cotton hairs on a slide in a drop of I_2KI reagent. After a minute, remove the I_2KI with a strip of filter paper and add a coverslip.

Apply a drop of 60% (v/v) sulfuric acid at the edge of the coverslip and observe the behavior of the fibers with a microscope. Notice the spiral movements that occur during the swelling process. Then press down on the coverslip until the pressure flattens the swollen fibers. Note and draw a sketch that illustrates the cellulose fibrils that make up the wall.

Set up a microscope slide with a few cotton hairs under a cover glass and add a drop of cuprammonia solution to the edge of the coverslip. Note the vigorous spiral movements of the fibers and the head-like swellings. Cellulose is rapidly dissolved by fresh cuprammonia, but the pectic compounds are insoluble in this reagent. Indicate the location of the pectic compounds in the cell walls of cotton hairs.

Observations:

B.   Detection of pectic compounds

Place some thin cross sections of the endosperm of a nasturtium seed (soaked for several hours in water) in a small dish containing a dilute solution of ruthenium red. After about twenty minutes (longer if the stain is not sharply differentiated), remove a section and mount it on a slide in a drop of water or glycerin. Examine the section with medium and high power. The pectic compounds stain red while the two other compounds are unstained or lightly stained. Draw two adjacent cells and indicate the position of the pectic compounds.

In a similar manner, stain a thin cross section of a young tomato stem with ruthenium red solution. Examine the collenchyma cell walls under medium and high power for the presence of pectic compounds. Illustrate where the pectic compounds are in the collenchyma walls. Also, note if other tissues in the section exhibit pectic compounds.

Observations:

C.   Detection of hemicelluloses

Mount a thin cross section of the endosperm of a soaked nasturtium seed on a slide in a drop of water and observe with a microscope. Note the irregular thickenings in the cell wall.

Mount another section of the endosperm on a slide in a drop of 12% HCl. Examine under medium power and note the thickness of the cell walls. Warm the slide over a flame for a few minutes and reexamine. Hemicelluloses are hydrolyzed by hot dilute mineral acids, whereas true cellulose is resistant to this treatment.

In the same manner as above, prepare thin cross sections of the endosperm of germinating nasturtium seeds and compare with those of ungerminated seeds.

Observations:

D.   Detection of lignin

Mount a thin cross section of a tomato and sunflower stem on a slide. Add a drop of alcoholic phloroglucinol and allow the solution to evaporate. Then add a drop of concentrated HCl and place a cover slip on the section. Lignified tissues stain cherry red. Indicate what tissues in the stem are lignified.

Observations:

E.   Detection of cutin

Place thin cross sections of *Aloe* or *Clivia* leaves in a small dish of Sudan III solution. After fifteen or twenty minutes remove a section, rinse it well in distilled water, and mount on a microscope slide in a drop of water. Examine the section under medium power of the microscope. Cutin stains orange-red. Draw the section examined and indicate the location of stained cutin.

Observations:

F.   Detection of suberin

Place thin cross sections of a geranium stem in a small dish of Sudan III solution. After fifteen or twenty minutes remove a section, wash it well in water, and mount it on a microscope slide in a drop of water. Examine the stained section under medium power of the microscope. Suberin, like cutin, stains orange-red. Indicate in a drawing where the suberin is located in the cell walls of this section.

Observations:

Results/
Conclusions

Present labeled sketches and/or concise statements to illustrate and explain the results. In concluding statements, consider the chemical structure of the cell wall components studied and the general basis for the tests used. Answer the following questions as a guideline to the interpretation of your results:

1.   What tissues in the stem are composed largely of cellulose?

2.   What tissues did not give indication of the presence of cellulose?

3.   If cellulose is not detected by I_2KI reagent, does this mean that cellulose is not present? Explain.

4.   On the basis of the test with the cuprammonium reagent, what can be inferred concerning the distribution of pectic compounds in the cell walls of cotton hairs?

5.   Where are pectic compounds located in the cell walls of nasturtium endosperm?

6.   Is pectin present in cell walls of tissues other than the collenchyma tissues of young tomato stems?

7.   With respect to hemicelluloses, what changes occur in endosperm cell walls during seed germination? What does this suggest regarding the role of hemicelluloses?

8.   What is the physiological significance of lignification?

9.   How can suberized cells be distinguished from cutinized cells? What is the physiological significance of cell walls containing cutin and suberin?

10.   What term is commonly applied to suberized cells?

Acknowledgment

This exercise was adapted with permission from B. S. Meyer, D. B. Anderson, and C. A. Swanson. 1955. *Laboratory Plant Physiology*, 3rd edition. D. Van Nostrand Co., Inc., Princeton, N.J., pp. 138–140.

References

Anderson, R. L. and B. A. Stone. 1978. Studies on *Lolium multiflorum* endosperm in tissue culture. III. Structural studies on the cell walls. *Aust. J. Biol. Sci.* 31: 573–586.

Brown, S. A. 1968. Lignins. *Ann. Rev. Plant Physiol.* 17: 223–244.

Fincher, G. B. 1976. Morphology and chemical composition of barley endosperm cell walls. *J. Inst. Brew.* 81: 116–122.

Fincher, G. B., B. A. Stone, and A. E. Clarke. 1983. Arabinogalactan-proteins: Structure, biosynthesis, and function. *Ann. Rev. Plant Physiol.* 34: 47–70.

Kolattukudy, P. E. 1981. Structure, biosynthesis, and biodegradation of cutin and suberin. *Ann. Rev. Plant Physiol.* 32: 539–567.

Labavitch, J. M. 1981. Cell wall turnover in plant development. *Ann. Rev. Plant Physiol.* 32: 385–406.

Lott, J. N. A., with J. T. Darley. 1976. *A Scanning Electron Microscope Study of Green Plants.* St. Louis: Mosby.

Mühlethaler, K. 1967. Ultrastructure and formation of plant cell walls. *Ann. Rev. Plant Physiol.* 18: 1–24.

Preston, R. D. 1979. Polysaccharide formation and cell wall function. *Ann. Rev. Plant Physiol.* 30: 55–78.

Detection of certain metabolic enzymes in tissues

Introduction

Many of the enzymes of cellular metabolism are found in the cytoplasm and organelles of cells. Compartmentation of enzymes takes place within the cell, which affords their association with substrates and efficient regulation of biochemical reactions. The compartmentation of enzymes reaches a high degree in cellular organelles, such as the mitochondria and chloroplasts. However, even the ground cytoplasm (cytoplasm exclusive of organelles) partitioned by the endoplasmic reticulum abounds in countless numbers of different enzymes and their respective substrates.

Some of the enzymes in plant cells may be detected by relatively simple chemical tests. The means of detection of common enzymes of plant cells are illustrated in the following experiments. Some of the more common enzymes easily detected include dehydrogenases, oxidase, peroxidases, and catalase.

Dehydrogenases catalyze reactions in which there is oxidation of a suitable substrate in which hydrogen ions and electrons are removed and used to reduce a suitable acceptor—namely, a coenzyme. For example, a well-known dehydrogenase, succinic acid dehydrogenase, is found in mitochrondia and mediates the dehydrogenation (oxidation) of succinic acid to fumaric acid. The two hydrogens and electrons removed from succinic acid reduce the coenzyme flavin adenine dinucleotide (FAD). To study dehydrogenases *in vitro*, a suitable dye, such as 2,3,5-triphenyltetrazolium chloride or methylene blue is usually reduced in place of the coenzyme. As the reduced acceptor is formed, there is a color change that may be measured colorimetrically to determine the extent of the reaction.

Oxidases are enzymes that mediate the removal of hydrogen or electrons from a substrate in which molecular oxygen is required for the reaction. To detect oxidases in plant tissue, a suitable phenolic (gum guaiacum) is often used to flood the tissue. The change in color of the tissue is usually indicative of the presence of an oxidase(s) enzyme.

Peroxidase activity is typified by the oxidation of phenols with hydrogen peroxide (H_2O_2) as the electron acceptor accordingly:

$$H_2O_2 \ + \ \text{phenol (reduced)} \xrightarrow{\text{peroxidase}} \text{phenol (oxidized)} + 2H_2O$$

The major distinction between peroxidases and oxidases is that a peroxidase reaction does not require the addition of oxygen. It should be noted that the chemical degradation of indole-3-acetic acid in plants may be due, at least in part, to peroxidative action. Peroxidases are detected in plant tissue by adding a suitable phenol to a plant extract, which undergoes a color change with oxidation.

Catalases provide reactions in which H_2O_2 is converted to water with the evolution of oxygen gas. The primary distinctions between catalase and peroxidase is that catalase does not require the presence of a phenol for the reaction with O_2 liberated in the process.

Purpose

To study some rapid techniques for the detection of dehydrogenase, oxidase, peroxidase, and catalase in plant tissues.

Materials

2 peeled white potato tubers

1 cauliflower head (15 g needed)

100 ml 2,3,5-triphenyltetrazolium chloride solution (0.5%, w/v)

100 ml methylene blue solution (0.025%, w/v)

100 ml gum guaiacum alcoholic solution (2%, w/v; 2 g of gum guaiacum dissolved in sufficient 95% ethanol to bring final volume to 100 ml)

hydrogen peroxide (commercial grade, 3%)

30 ml hydrogen peroxide (1 part 3% hydrogen peroxide to 30 parts water)

200 ml ethanol (95%)

200 ml citrate buffer: 10 g citric acid (monohydrate) and 95.3 ml 1N NaOH mixed with an equal volume of water (approximately pH 5.0)

25 ml pyrogallol solution (5%, w/v)

ethyl ether

10 ml sulfuric acid (10%)

4 petri dishes

2 test tubes

boiling water bath

mortar and pestle

quartz sand

filter paper

Buchner funnel and side-arm suction flask

graduated cylinder (100 ml)

2 Erlenmeyer flasks (250 ml)

separatory funnel

Procedure   **A. Dehydrogenase**

Prepare several slices of potato tuber tissue. Transfer half of the slices to a small dish and cover with 0.5% 2,3,5-triphenyltetrazolium chloride solution. Observe the red color that develops as the compound is reduced. The section that indicates the presence of dehydrogenase enzymes usually occurs within fifteen minutes.

Immerse the remaining slices in boiling water for at least five minutes. After this time, remove the slices and immerse them in the dye solution as before.

Observations:

An alternative procedure for detecting dehydrogenase may be performed by cutting ten small cubes of potato tissue (5 mm on an edge) from a potato tuber. Transfer five of the cubes to a small test tube and place the other five in a small beaker of boiling water and allow them to remain at least twenty minutes. Then transfer them to a small test tube.

Fill each tube completely with a solution of 0.025% methylene blue and stopper tightly. Eliminate all air bubbles within the tubes. At the end of twenty-four hours' incubation, observe any color change in the tubes.

Remove the cubes from the tubes and lay them on a piece of paper so that they are exposed to air. Observe the color change that occurs.

Observations:

B.   Oxidase

Cut a thin transverse section of a potato tuber including at least one bud section. Place the slice in a petri dish and flood its surface with a freshly prepared 2% alcoholic solution of gum guaiacum. The development of blue color resulting from the oxidation of guaiacum usually occurs within fifteen minutes and indicates the presence of an oxidase enzyme. Repeat the test with a potato slice that has first been immersed in boiling water for at least five minutes.

Observations:

C.   Peroxidase

From potato tissue

Prepare a thin transverse section of a potato tuber including at least one bud section. Place in a petri dish and flood the surface of the slice with a freshly prepared 2% alcoholic solution of gum guaiacum. After ten to fifteen minutes, drain the guaiacum solution from the surface of the slice and replace with a dilute solution of hydrogen peroxide (1 part 3% hydrogen peroxide to 30 parts water). Note any differences in the rapidity and intensity of the development of the blue color as compared with the oxidase test. Repeat the test with a potato slice that has first been immersed in boiling water for at least five minutes.

Observations:

From cauliflower

1.   Peroxidase solution: Grind 15 g of cauliflower in a mortar with a pinch of quartz sand and 20 ml of 95% alcohol for two minutes. Add 70 ml of alcohol and grind further. Then collect the crude fiber on filter paper in a Buchner funnel and transfer it to a 100 ml graduated cylinder. Resuspend in a sufficient amount of citrate buffer (pH 5.0) to bring the final volume to 100 ml. Stir the powder into the buffer and incubate for about one hour.

2.   Reaction mixture and purpurogallin formation: Transfer 20 ml of the peroxidase preparation into a test tube. Place the tube in boiling water and leave there for at least fifteen minutes. Transfer another 20 ml protein of the peroxidase preparation to a 250 ml flask containing 180 ml of distilled water. Then add 10 ml of *freshly prepared* 5% pyrogallol solution and 1 ml of 3% hydrogen peroxide. Mix well and after five minutes add 5 ml of 10% H_2SO_4 and mix again.

Extract the amber-colored purpurogallin formed by shaking out into ether, using a separatory funnel. Repeat the procedure using the boiled peroxidase preparation. Determine the extent of the reaction in both cases by comparing the relative amounts of purpurogallin formed.

Observations:

D.   Catalase

Cover a slice of potato tuber with dilute hydrogen peroxide (1 part 3% H_2O_2 to 30 parts water). Observe the evolution of oxygen bubbles, which denote the presence of the enzyme. Repeat with a potato slice that has first been immersed in boiling water for at least five minutes.

Observations:

Results/
Conclusions

Indicate the basic reaction sequence involved in each test and the biological importance of the enzyme studied.

Acknowledgment

This exercise was adapted with permission from B. S. Meyer, D. B. Anderson, and C. A. Swanson. 1955. *Laboratory Plant Physiology*, 3rd edition. D. Van Nostrand Co., Inc., Princeton, N.J., pp. 110–112.

References

Glick, D. 1949. *Techniques of Histo- and Cyto-Chemistry*. New York: Interscience Publications, Inc.

Krebs, H. A. 1970. The history of the tricarboxylic acid cycle. *Perspectives in Biol. and Med.* 14: 154–170.

Lehninger, A. L. 1982. *Principles of Biochemistry*. New York: Worth.

Styles, W. and W. Leach. 1960. *Respiration in Plants*, 4th edition. New York: John Wiley and Sons.

Stryer, L. 1981. *Biochemistry*, 2nd edition. San Francisco: W. H. Freeman and Co.

EXERCISE 34

Succinic acid dehydrogenase activity of plant mitochondria

Succinic acid dehydrogenase, one of the numerous enzymes that play an important role in the operation of the Krebs cycle (citric acid cycle, tricarboxylic acid cycle), is found in mitochondria. As mentioned previously (see the introduction to Exercise 33), it mediates the dehydrogenation (oxidation) of succinic acid. Interestingly, this reaction appears to be the only Krebs cycle oxidation that does not employ a pyridine nucleotide (NAD). Instead, succinic acid is oxidized with the reduction of the flavin prosthetic group (coenzyme), flavin adenine dinucleotide (FAD). The oxidation of succinic acid represents the third oxidation step of the Krebs cycle with the formation of furmaric acid.

For studies relating to the action of succinic acid dehydrogenase, mitochondria are first isolated from a suitable source such as cauliflower heads (florets). A source of plant material devoid of pigmentation is often selected to eliminate the interfering influence of pigments on the isolation of mitochondria and on the colorimetric assay of the enzyme. Further, as is the case with natural product isolation, particularly enzymes, the material being isolated should be kept cold.

For the following exercise, mitochondria are isolated by high-speed centrifugation according to the procedures outlined. The relatively pure mitochondrial preparation obtained is then incorporated into a suitable reaction mixture containing the necessary reaction components, including the hydrogen acceptor 2,6-dichlorophenol-indophenol (DCIP). Potassium cyanide (KCN)—*use caution*—is also used to block the electron transport system (cytochromes) so that high levels of oxidized, naturally occurring acceptors (FAD) are minimal.

As the reaction proceeds, the blue 2,6-DCIP (oxidized form) should become increasingly lighter in color or colorless (reduced form). The extent of the reaction is then evaluated colorimetrically. The procedures for mitochondrial isolation and assay for succinic acid dehydrogenase are outlined in the following experiments.

Purpose

To assay for succinic acid dehydrogenase activity in a mitochondrial preparation from cauliflower.

Materials

1 head of cauliflower (*Brassica oleracea* L.)

45 ml grinding solution consisting of 0.5M sucrose and 0.1M KH_2PO_4 (171 g sucrose and 13.6 g KH_2PO_4 dissolved and diluted to 1 liter with water and adjusted to pH 7.3)

4 ml suspending solution (pH 7.2) containing 0.4M sucrose, 0.02M KH_2PO_4, 0.005M $MgCl_2$, and 0.02M glucose (136.8 g sucrose, 2.72 g KH_2PO_4, 0.475 g $MgCl_2$, and 3.60 g glucose dissolved and diluted to 1 liter with water)

40 ml phosphate buffer, pH 7.0 (see Appendix II)

10 ml 2,6-dichlorophenol indophenol solution (1.7 × 10^{-3}M)

5 ml $MgSO_4$ (0.5M)

5 ml KCN (0.1M)

5 ml HCl (5%, v/v)

10 ml sodium succinate (0.25M), adjusted to pH 7.0 with NaOH

paper towels

mortar and pestle

razor blades

quartz sand

cheesecloth (two layers)

refrigerated centrifuge, centrifuge tubes

glass tissue homogenizer

pipets (1 ml, 5 ml graduated)

cuvettes (or Klett tubes)

spectrophotometer or colorimeter (set at 620 nm)

balance

Procedure

A. Mitochondrial preparation from cauliflower

Cauliflower heads are an excellent source of mitochondria. Before using, wash the heads in tap water, rinse in distilled water, wrap in a damp towel, and cool in a refrigerator preferably twelve hours before the laboratory meeting. In addition, all solutions and containers used for the mitochondrial preparation should be ice-cold.

Remove the top 2 to 3 mm of the immature inflorescences (florets) and discard the remainder of the head. Complete the extraction according to the following steps (preparation of the mitochondrial fraction may be completed before the class meeting):

1. Grind approximately 15 g of cauliflower florets in a chilled mortar with 35 ml of cold grinding solution (0.5M sucrose plus 0.1M KH_2PO_4, pH 7.3). A small quantity of cold reagent-grade white sand may be sprinkled on the tissue to facilitate grinding.

2. Pass the suspension through a layer of cheesecloth and centrifuge the filtrate at 500 × g for ten minutes at 4°C.

3. Decant the supernatant and centrifuge it for thirty minutes at 20,000 × g (0 to 4°C).

4. Discard the supernatant and resuspend the pellet in 10 ml of grinding solution. Use an electric homogenizer, if available.

5. Recentrifuge the suspension (from step 4) for thirty minutes at 20,000 × g (0 to 4°C) and discard the supernatant. (This step is optional depending on the relative purity desired and time available.)

6. Suspend the pellet (containing the mitochondria) in 4 ml of suspending solution (0.4M sucrose; 0.02M KH_2PO_4, pH 7.2; 0.02M glucose; 0.005M $MgCl_2$). The preparation may be frozen at once and used later, but immediate use is recommended for optimum enzyme activity.

B. Assay of enzyme activity

To perform a simple assay for succinic acid dehydrogenase activity, add the stock solutions in milliliters (Table 34–1) to standard Klett or colorimeter tubes numbered 1 through 7. The amount of substrate, mitochondrial preparation, or the dye to be used may have to be slightly altered to give optimum results. However, use Table 34–1 as a guide and then make suitable adjustments as deemed necessary.

TABLE 34–1   Reaction mixtures for succinic acid dehydrogenase activity

Stock Solutions*	Tube Number						
	1	2	3	4	5	6	7
Phosphate buffer (pH 7.0)	1.0	1.0	1.0	1.0	1.0	1.0	1.0
MgSO₄ (0.5M)	0.1	0.1	0.1	0.1	0.1	0.1	0.1
KCN (0.1M)	0.2	0.2	0.2	0.2	0.2	0.2	0.2
HCl (5%, v/v)	0.1	0.1	0.1	0.1	0.1	0.1	0.1
2,6-DCIP (1.7 × 10⁻³M)	None	0.2	0.2	0.2	0.2	None	0.2
Water	3.8	4.0	3.6	4.0	3.6	3.6	3.4
Mitochondrial preparation	0.4	None	0.4 (Boiled)	0.4	0.4	0.6	0.6
Sodium succinate (0.25M)	0.4	0.4	0.4	None	0.4	0.4	0.4
Total volume	6.0	6.0	6.0	6.0	6.0	6.0	6.0

* The numbers indicated represent the amount in millions of the stock solution to be used.

Add the substrate last and when the reactions are to be started, mix the contents by agitation of each test tube. Use tubes 1 and 2 as blanks in the spectrophotometer or colorimeter. In addition, tubes 1 and 6 should be used to adjust for the varying amounts of turbidity given by the different amounts of mitochondrial preparation in tubes 5 and 7, respectively. Record the changes in percent transmittance of the reaction mixtures at 620 nm at 0 time, and at ten-minute intervals for sixty minutes. In the test tubes containing the complete reaction mixture, the blue 2,6-DCIP (oxidized form) should become increasingly lighter in color or colorless (reduced form). The extent of the reaction can be evaluated by comparison of all tubes against the controls. Record your results in Table 34–2.

TABLE 34–2   Results of succinic acid dehydrogenase activity with time

Tube Number	Time in Minutes						
	0	10	20	30	40	50	60
	Percent Transmittance at 620 nm						
1							
2							
3							
4							
5							
6							
7							

Depending upon the amount of time and facilities available, you may perform various experiments to determine reaction rates on the basis of increasing substrate concentrations (with enzyme constant), increasing enzyme concentration (with substrate constant), or the effect of a known competitive inhibitor (malonic acid). Experiments are left to the discretion of the instructor.

Results/ Conclusions

Record your results of enzyme activity with time in Table 34–2. For the additional suggested experiments, present a graph in which enzyme activity is plotted on the ordinate against the variable being tested.

Conclusions may be presented concerning the details of the mitochondrial isolation procedure and the succinic acid dehydrogenase mediated reaction. Be sure to consider the nature of the substrate and product and the role of other components used in the reaction mixtures. Name the other enzymes or group of enzymes present in mitochondria. In addition, consider the structure of mitochondria and their general function.

References

Bonner, W. D., Jr. 1973. Mitochondria and plant respiration. In Miller, L. P. (ed.). *Phytochemistry*. New York: Van Nostrand Reinhold.

Cooperstein, S. J., A. Lazarow, and N. J. Kurfess. 1950. A microspectrophotometric method for the determination of succinic dehydrogenase. *J. Biol. Chem.* 186(1): 129–139.

Ikuma, H. 1972. Electron transport in plant respiration. *Ann. Rev. Plant Physiol.* 23: 419–436.

Krebs, H. A. 1970. The history of the tricarboxylic acid cycle. *Perspectives in Biol. and Med.* 14: 154–170.

Lehninger, A. L. 1982. *Principles of Biochemistry*. New York: Worth.

Mitchell, P. 1966. Chemiosmotic coupling in oxidative and photosynthetic phosphorylation. *Biol. Rev.* 41: 445–502.

EXERCISE 35

Oxygen uptake measured by a polarograph

Introduction

Introduction

Mitochondria are bounded by two-unit membranes that enclose an inner matrix. Numerous folds of the inner membrane (termed cristae) project into this matrix. Analyses of mitochondria reveal the presence of phospholipids, DNA and RNA, the Krebs cycle enzymes and their substrates, the cytochromes, and components of the electron transport system (ETS).

Mitochondria provide a great deal of the cell's usable energy in the form of high-energy phosphate compounds, primarily ATP. The production of ATP through the oxidation of Krebs cycle intermediates and coupling with the electron transport system is referred to as *oxidative phosphorylation*. Molecular oxygen is required for the operation of the ETS because it is the terminal acceptor of electrons and necessary for the ultimate formation of water. The flow of electrons (oxidation-reduction reactions of the ETS) generates the necessary energy to produce ATP. However, without oxygen, the ETS system will not operate because in the absence of the terminal acceptor (oxygen), the cytochromes will remain in the reduced state and not accept additional electrons. Thus mitochondria are the organelles responsible for aerobic respiration and take up oxygen in response to the levels of reduced components of the electron transport system. A measure of the respiratory activities of mitochondria or tissues in which they are found is related to the amount of oxygen uptake.

In the past, one means of determining the respiratory activities of mitochondria, plant cells, or tissue involved the use of respirometers (Warburg or Gilson) to measure O_2 uptake and CO_2 evolution by plants, tissues, or mitochondrial preparations. One simple method currently used is to assess the amount of oxygen taken up by plant tissues, cells, or mitochondria from the surrounding medium by means of an oxygen electrode, which is used to measure the disappearance of oxygen as respiration proceeds. The following experiment is designed to demonstrate the use of the oxygen electrode in measuring oxygen uptake by isolated mitochondria. Further details relating to the reactions of the Krebs cycle and ETS are outlined in the texts cited at the end of the exercise.

Purpose

To become familiar with the Clark oxygen polarograph method of measuring oxygen uptake by cauliflower mitochondria.

Materials

1 head of cauliflower

grinding solution consisting of 0.5M sucrose and 0.1M KH_2PO_4 (171 g sucrose and 13.6 g KH_2PO_4 dissolved and diluted to 1 liter with water and adjusted to pH 7.3)

suspending solution (pH 7.2) containing 0.4M sucrose, 0.02M KH_2PO_4, 0.005M Mg_2Cl, 0.02M glucose (136.8 g sucrose, 2.72 g KH_2PO_4, 0.475 g $MgCl_2$, and 36 g glucose dissolved and diluted to 1 liter with water)

Clark oxygen polarograph, such as Yellow Springs Instruments

YSI 5533

recorder for above instrument (optional)

replacement membrane material and kit

disposable micropipets

mortar and pestle

razor blade

cheesecloth

centrifuge (high-speed)

centrifuge tubes to fit centrifuge

homogenizer (glass or teflon) and motor

Erlenmeyer flask (125 ml)

Procedure

A. Mitochondrial preparation from cauliflower

Cauliflower heads are an excellent source of mitochondria. Before using, wash the heads in tap water, rinse in distilled water, wrap in a damp towel, and cool in a refrigerator preferably twelve hours before the laboratory meeting. In addition, all solutions and containers used for the mitochondrial preparation should be ice-cold.

Remove the top 2 to 3 mm of the immature inflorescences (florets) and discard the remainder of the head. Complete the extraction according to the following steps (preparation of the mitochondrial fraction may be completed before the class meeting):

1. Grind approximately 15 g of cauliflower florets in a chilled mortar with 35 ml of cold grinding solution (0.5M sucrose plus 0.1M KH_2PO_4, pH 7.3). A small quantity of cold reagent-grade white sand may be sprinkled on the tissue to facilitate grinding.

2. Pass the suspension through a layer of cheesecloth and centrifuge the filtrate at 500 × g for ten minutes at 4°C.

3. Decant the supernatant and centrifuge it for thirty minutes at 20,000 × g (0 to 4°C).

4. Discard the supernatant and resuspend the pellet in 10 ml of grinding solution. Use an electric homogenizer, if available.

5.   Recentrifuge the suspension (from step 4) for thirty minutes at 20,000 × g (0 to 4°C) and discard the supernatant. (This step is optional depending on the relative purity desired and time available.)

6.   Suspend the pellet (containing the mitochondria) in 4 ml of suspending solution (0.4M sucrose; 0.02M KH_2PO_4, pH 7.2; 0.2M glucose; 0.005M $MgCl_2$).

B.   Enzyme activity and oxygen uptake

Pipet 4.4 ml of suspending medium into one vial. Saturate the suspending medium with oxygen before adding to the vial by stirring in a flask.

Place the probe with membrane intact into the vial carefully to avoid air bubbles in the medium. Turn the instrument on to zero setting and be certain the stirring bar is revolving freely. Set meter dial on zero. Switch to "read" and with the calibrator dial set to 100%, after stability of the instrument is ascertained, add 0.6 ml of mitochondrial preparation (more may be needed depending on the preparation). Begin to take meter readings and/or record the results. Continue for ten minutes (or until near zero O_2 uptake is attained), then carefully pipet 60 μl of 1M sodium succinate into the reaction vial and begin taking readings at that point. Continue taking readings until the instructor advises you to stop. Record the data in tabular form.

Instrument readings:

Results/
Conclusions

In concluding statements, answer the following:

1.   How is the structure of the mitochondrion related to its function?

2.   What is meant by the term *endogenous respiration*?

3.   How does the electron transport system (ETS) in mitochondria relate to the uptake of oxygen by these organelles?

4.   Why did the readings change when sodium succinate was added to the reaction mixture?

5.   What is meant by the term *coupled reactions*?

References

Goodwin, T. W. and E. I. Mercer. 1982. *Introduction to Plant Biochemistry*, 2nd edition. New York: Pergamon Press.

Hall, J. L, T. J. Flowers, and R. M. Roberts. 1974. *Plant Cell Structure and Metabolism*. New York and London: Longman, Limited.

Lehninger, A. L. 1982. *Principles of Biochemistry*. New York: Worth.

Packer, Lester. 1967. *Experiments in Cell Physiology*. New York: Academic Press.

Theologis, A. 1979. The genesis development and participation of cyanide-resistant respiration in plant tissue. Ph.D. Thesis, University of California, Los Angeles.

Umbreit, W. W., R. H. Burris, and J. F. Stauffer. 1972. *Manometric and Biochemical Techniques*, 5th edition. Minneapolis: Burgess Publishing Co.

White, A., P. Handler, E. L. Smith, R. L. Hill, and I. R. Lehman. 1978. *Principles of Biochemistry*, 6th edition. New York: McGraw-Hill.

EXERCISE 36

The extraction and properties of phenylalanine deaminase (L-phenylalanine ammonia lyase)

Introduction

The enzyme 6-phenylalanine ammonia lyase mediates the conversion of L-phenylalanine to *trans*-cinnamic acid. This enzyme is widespread in plants.

During the course of the following exercise, L-phenylalanine ammonia lyase will be extracted from four-day-old *Avena* seedling shoots (stems and leaves) and isolated as an acetone powder for enzyme assays. Using the enzyme preparation, the basic assay will be performed with emphasis on techniques for conducting a time course study, a study on the pH optimum for enzyme activity, the effect of substrate concentration on enzyme activity, and the effect of enzyme concentration in the reaction. The mentioned studies are important for characterizing enzymes and showing certain factors that affect their activity.

The reaction mixture necessary for assaying the enzyme contains the substrate phenylalanine, water, borate buffer (pH 8.8), and enzyme preparation. After the reaction is stopped by the introduction of HCl (5N), the amount of *trans*-cinnamic acid is determined by the absorbance of the reaction mixture at 290 nm. The product *trans*-cinnamic acid strongly absorbs, while the substrate, phenylaline, has almost no absorption at 290 nm.

While performing the experiments, pay special attention to the techniques of pipeting, the operation of the spectrophotometer, and the techniques outlined.

Purpose

To extract phenylalanine deaminase from four-day-old *Avena* seedlings and to study the effect of pH, substrate, enzyme, and time on the enzymatic conversion of L-phenylalanine to *trans*-cinnamic acid.

164

Materials

40 g four-day-old *Avena* seedlings
(see Procedure, Part A)

buffers (see Appendix II)

17 ml enzyme preparation (see
Procedure, Parts C and D)

25 ml phenylalanine solution (10
mg/ml H$_2$O, w/v)

5 ml phenylalanine solution (5
mg/ml H$_2$O, w/v)

100 ml HCl (5N)

1000 ml acetone (−15°C)

analytical balance

Buchner funnel (attached to a
suction pump)

filter paper (Whatman No. 3)

mortar and pestle (cold)

ice bath

water bath (30°C)

Waring blender

pH meter

centrifuge tubes

centrifuge (refrigerated)

magnetic stirrer (or stirring rods)

33 test tubes

cuvettes (matched)

ultraviolet spectrophotometer (set
at 290 nm)

Procedure

A.   Four-day-old *Avena* seedlings

Soak *Avena* seeds for thirty minutes in a 0.05% solution of Tween-20. Then plant seeds in vermiculite and maintain in darkness for four days. Forty g of plant material requires about 800 seedlings (two flats) and will give about 2 g of acetone powder containing the enzyme. The amount of plant material, therefore, may be scaled up or down accordingly.

B.   Enzyme extraction and preparation of acetone powder

Homogenize approximately 40 g (fresh weight) of four-day-old *Avena* seedlings (use the shoots harvested by cutting at the level of vermiculite) in 300 ml of cold acetone (−15°C) for sixty seconds in a Waring blender.

Using suction, filter the homogenate through a Buchner funnel (lined with one sheet of Whatman No. 3 filter paper). Wash the residue remaining on the filter paper several times with additional volumes of cold acetone. Spread the residue and allow it to dry at room temperature. Powder the dried residue with a mortar and pestle. Store in a closed container in a freezer. The acetone powder should be stable for at least several weeks when stored in the freezer.

C.   Enzyme preparation for reaction mixtures

Suspend the acetone powder in cold borate buffer (buffer No. 1, pH 8.8) at a concentration of 10 mg/ml buffer. Be sure to make up sufficient enzyme preparation for the following experiments.

Maintain the suspension at 4°C for thirty minutes with constant but gentle agitation. Use a magnetic stirrer.

Centrifuge the suspension at 14,000 × g in a refrigerated centrifuge for twenty to thirty minutes. The *supernatant* that contains the enzyme should be carefully decanted and used *immediately* or frozen. The activity in the frozen enzyme preparation can be maintained for a few days. Therefore, it is suggested that the following experiments be planned accordingly so that you will not have to prepare the enzyme too often.

D. Phenylalanine deaminase assay

To perform a simple assay for enzyme activity, label three test tubes and add the components as indicated in Table 36–1.

TABLE 36–1   Phenylalanine deaminase activity

Components	Tube Number and Volume Added (ml)		
	1	2	3
Phenylalanine (10 mg/ml H$_2$O)	0	1.0	1.0
Distilled water	3.5	2.5	2.5
Borate buffer (0.1M, pH 8.8)	1.5	1.5	1.5
Enzyme preparation*	0.5	0.5 (Boiled)	0.5
Total volume	5.5	5.5	5.5

* Add enzyme to start reaction and incubate at 30°C for one hour.

After the addition of the last component (enzyme preparation), shake the tubes gently and incubate the mixtures at 30°C for one hour. At the end of incubation time, stop the reaction by the addition of 0.5 ml of 5N HCl to each tube. Then transfer the contents of each tube to a cuvette and measure the absorbancy at 290 nm in a suitable spectrophotometer. Record the optical density readings.

The extinction coefficient of the product, *trans*-cinnamic acid, dissolved in the borate buffer at pH 8.8, is slightly more than 10,000 at 290 nm. On the other hand, the substrate phenylalanine has almost no absorption at this wavelength.

To determine the amount of enzyme activity on the basis of absorbency, subtract the optical density (OD) value obtained for the reaction mixture containing boiled enzyme (tube 2) from that of the mixture containing active enzyme (tube 3). Now using the difference, estimate the amount of enzyme activity on the basis of enzyme units. An enzyme unit is defined as that amount of enzyme that produces an increase in absorbency of 0.01 per hour in the standard assay. In addition, this change in absorbency is equivalent to the formation of approximately 1 µg of cinnamic acid per 6 ml of the final mixture (that is, reaction mixture; 5.5 ml plus 0.5 ml 5N HCl). Therefore, you should be able to calculate the amount of cinnamic acid formed in the complete reaction mixture.

Optical density values and calculations:

E.   Time course study

Add the following components to each of six test tubes (numbered 2 through 7):

Components	Volume (ml)
Phenylalanine (10 mg/ml H_2O)	1.0
Distilled water	2.8
Borate buffer (pH 8.8)	1.5
Enzyme preparation (add to start reaction)	0.5
Total volume	5.5

Do not add enzyme until the reactions are to be started. Set up an additional tube (tube 1) in which the substrate is omitted and in which the total volume is adjusted to 5.5 ml with additional water.

To start the reactions (at time 0) add the enzyme (0.5 ml) to each tube. Then add 0.5 ml of 5N HCl to tube 1 *only*. For the remaining tubes, stop the reactions at ten-minute intervals (over the course of one hour) by the addition of 0.5 ml of 5N HCl. For example, the reaction in tube 2 should be stopped ten minutes after time 0; tube 3, twenty minutes; tube 4, thirty minutes, and so on.

After stopping the reaction in each tube, adjust the spectrophotometer with a blank at 290 nm and determine the absorbency for the appropriate reaction mixture. Record the optical density value obtained for each reaction mixture.

Optical density values and amount of product formed:

F.   The pH optimum for enzyme activity

Label six test tubes and add the various components indicated in Table 36–2. Follow the same procedures outlined previously for starting and stopping the reactions and for assaying enzyme activity. Since the optimum pH for enzyme activity is to be determined on a relative basis, it is not essential to run a complete reaction mixture containing inactivated enzyme. Therefore, the optical density values as measured directly may be used to calculate the amount of end product formed at different pH levels.

TABLE 36–2   The pH optimum for enzyme activity

Components	Tube Number and Volume Added (ml)					
	1*	2	3	4	5	6
Phenylalanine (10 mg/ml H_2O)	0	1.0	1.0	1.0	1.0	1.0
Distilled water	3.5	2.5	2.5	2.5	2.5	2.5
Phosphate buffer (pH 6.2)	0	1.5	0	0	0	0
Borate buffer (pH 8.0)	0	0	1.5	0	0	0
Borate buffer (pH 8.8)	1.5	0	0	1.5	0	0
Borate buffer (pH 9.0)	0	0	0	0	1.5	0
Borate buffer (pH 10.0)	0	0	0	0	0	1.5
Enzyme preparation†	0.5	0.5	0.5	0.5	0.5	0.5
Total volume	5.5	5.5	5.5	5.5	5.5	5.5

* Reagent blank.
† Add the enzyme to start reaction and incubate at 30°C for one hour.

Optical density readings and calculations:

G.   Effect of substrate concentration on enzyme activity

Using the same components and the same procedures outlined previously, perform an experiment in which only the substrate is varied.

A good range of substrate concentrations to work with is as follows: 0, 0.5, 1, 1.5, 2, 3, and 5 mg of phenylalanine/reaction mixture. The mixture with no substrate may be used as the reagent blank.

Remember that it is absolutely essential to keep the same concentration of the enzyme, the buffer (use borate buffer No. 1, pH 8.8), and total volume (5.5 ml) for all the reaction mixtures. One way to vary the substrate concentration in accordance with the above is first to prepare a solution of phenylalanine at a concentration of 5 mg/ml of water. Then add the appropriate amounts to the respective tubes. The final volume of the reaction mixture may then be adjusted to 5.5 ml by the addition of an appropriate amount of distilled water.

In the space provided, present your experimental design in the form of a table. Then perform the experiment according to your outline.

Experimental design and results:

H.   Effect of increasing enzyme concentration

Design and perform an experiment that tests the effect of increasing enzyme concentration on the rate of conversion of phenylalanine to cinnamic acid. Follow the same basic scheme as before, but this time use different amounts of the enzyme preparation.

An easy method to follow is to incorporate the enzyme preparation (10 mg acetone powder/1 ml buffer) into the respective reaction mixtures according to the following values: 0, 0.2, 0.4, 0.5, 0.7, 0.9, 1 ml. All reaction mixtures should contain: 1 ml substrate (10 mg/ml H_2O), 2.5 ml H_2O, the different amounts of enzyme preparation, and sufficient borate buffer (pH 8.8) to adjust the final volume to 5.5 ml. Also set up a reagent blank (no substrate). Remember that an appropriate blank will be required for each amount of enzyme to adjust for changes in turbidity.

In the space provided, present your experimental design and the results of your experiment.

Experimental design and results:

Results/
Conclusions

For each part (E, F, G, H) present a graph in which enzyme activity (based on optical density changes or the amount of product formed) is plotted as the ordinate against the variable tested (that is, substrate concentration, enzyme concentration, pH, and time) on the abscissa.

Interpret the results on the basis of the various factors that influence enzyme activity. Include the reasons why a particular factor influences enzyme activity. Also, present a conventional chemical equation (show the structures of the substrate and product) that illustrates the enzymatic conversion of phenylalanine to *trans*-cinnamic acid.

With respect to higher plants, what is the importance of this reaction as it relates to the ultimate synthesis of various flavonoids (anthocyanins, and so on) and various compounds, such as caffeic acid, *p*-coumaric acid, quinnic acid, and chlorogenic acid? Are any of the compounds mentioned important in the regulation of plant growth and development? Explain.

Acknowledgment

The techniques of this exercise were adapted from the work of M. Zucker (see the references that follow) by W. G. Hopkins, Bryn Mawr College.

References

Hanson, K. R. and M. Zucker. 1963. The biosynthesis of chlorogenic acid and related conjugates of the hydroxycinnamic acids. Chromatographic separation and characterization. *J. Biol. Chem.* 238: 1105–1115.

Havir, E. A. and K. R. Hanson. 1968. L-Phenylalanine ammonia-lyase. I. Purification and molecular size of the enzyme from potato tubers. *Biochem.* 7: 1896–1903.

Havir, E. A. and K. R. Hanson. 1968. L-Phenylalanine ammonia-lyase. II. Mechanism and kinetic properties of the enzyme from potato tubers. *Biochem.* 7: 1904–1914.

Koukol, J. and E. E. Conn. 1961. The metabolism of aromatic compounds in higher plants. IV. Purification and properties of the phenylalanine deaminase of *Hordeum vulgare. J. Biol. Chem.* 236: 2692–2698.

Marsh, H. V., Jr., E. A. Havir, and K. R. Hanson. 1968. L-Phenylalanine ammonia-lyase. III. Properties of the enzyme from maize seedlings. *Biochem.* 7: 1915.

Zucker, M. 1965. Induction of phenylalanine deaminase by light and its relation to chlorogenic acid synthesis in potato tuber tissue. *Plant Physiol.* 40: 779–784.

The effect of light and temperature on photosynthesis

A technique used in the past to observe the evolution of oxygen gas from a photo-synthesizing plant entails the immersion of a sprig of an aquatic plant (for example, *Elodea*) in a solution of sodium bicarbonate. Counting the bubbles released from the sprig under different conditions provides a means for observing the relative effects of different environmental conditions—primarily light intensity and quality—on the rate of photosynthesis. Although this method is no longer used, it is included here for historical reasons and to provide students with an appreciation of the techniques used in the absence of advanced instruments and facilities.

Purpose

To determine the effect of light intensity and temperature with time on the rate of photosynthesis in *Elodea* sprigs.

Materials

sprigs of *Elodea*
100 ml sodium bicarbonate solution (0.1%)
large test tube (about 18 mm in diameter)
razor blade
glass rod

source of high-intensity white light
light meter (if available)
ruler
thermometer (C°)
beaker

Procedure

A.   Bubble counting method

Take a healthy sprig of *Elodea* and cut off the lower end. Immerse the apical end downward in a large test tube containing an aqueous solution of sodium bicarbonate (0.1%). If the cut end does not stay under the solution, attach the sprig to a glass rod to hold it under.

171

B.   Relative rate of photosynthesis under different light conditions

With the solution in the test tube at room temperature, expose the *Elodea* sprig to strong artificial light for two minutes, then count the number of bubbles given off during three periods of one minute each. Now place the plant in diffused light for two minutes and then determine the rate of photosynthesis as before. If the sprig does not respond adequately, select another and repeat the above. Place the plant in the darkest part of the room and after two minutes determine the rate again. By several trials determine how quickly the plant will respond to an abrupt change in light intensity.

This experiment can be refined in many ways. The relation between light intensity and distance between the object and the light source is as follows:

$$I_2 \ = \ I_1 \left(\frac{d_1}{d_2}\right)^2$$

Thus, if the original distance (d_1) between the light and the plant is doubled (d_2), the original intensity (I_1) is reduced by a factor of 4 (new intensity $= I_2$). Using this equation and an original light intensity reading (use a photocell if available), gradually increase the distance of the plant from the light source (nearest distance in cm) and record the photosynthetic rate over several one-minute determinations. Be sure the temperature of the sodium bicarbonate solution in which the *Elodea* sprig is immersed is kept constant. If desired, a photocell can be used to measure each new light intensity, but this is not absolutely necessary because the same relationship between distance from the light source and the photosynthetic rate may be determined.

Observations:

C.   Effect of temperature on rate of photosynthesis in *Elodea*

Determine the rate of photosynthesis in bright light at 20, 30, 40, and 50°C. Keep the light intensity constant and for each temperature make ten one-minute counts. To raise the temperature, transfer the test tube to a beaker of water previously heated to one or two degrees above the desired temperature.

Results/
Conclusions

Plot curves showing the effects of light intensity and temperature in relation to time on the rate of photosynthesis. In evaluating the results, you should consider the following:

1. The gas released from the plant
2. The method used in estimating the rate of photosynthesis
3. The units used in measuring light intensity
4. The difference between light intensity and amount of light
5. The reason for using a sodium bicarbonate solution

Give several possible reasons why no bubbles are released at low light intensities and under what conditions variations in light intensity might better affect photosynthesis.

Describe a situation where increasing the carbon dioxide concentration would have no effect on the rate of photosynthesis. Also list all the factors considered necessary for photosynthesis and discuss how the factors limit the process in a normal plant over a twenty-four-hour period (day and night).

References

Anderson, J. M. 1975. The molecular organization of chloroplast thylakoids. *Biochem. Biophys. Acta* 416: 191–235.

Arnon, D. I. 1971. The light reactions of photosynthesis. *Proc. Nat. Acad. Sci. U.S.* 68: 2883–2892.

Becker, W. M. 1977. *Energy and the Living Cell: An Introduction to Bioenergetics.* New York: Harper & Row Publishers, Inc.

Cogdell, R. J. 1983. Photosynthetic reaction centers. *Ann. Rev. Plant Physiol.* 34: 21–45.

Karp, G. 1984. *Cell Biology*, 2nd edition. New York: McGraw-Hill Book Co.

Malkin, R. 1982. Photosystem I. *Ann. Rev. Plant Physiol.* 33: 455–479.

San Pietro, A., F. A. Greer, and T. J. Arny (eds.). 1967. *Harvesting the Sun: Photosynthesis in Plant Life.* New York: Academic Press.

EXERCISE 38

Measurement of photosynthesis in leaf disks

Introduction

Another relatively straightforward technique used in the past for studying the effect of light intensity on the rate of photosynthesis is based on the evacuation of gas from leaf disks and then infiltration with sodium bicarbonate.

Nonphotosynthesizing leaf disks, when properly evacuated, will sink to the bottom of a solution of sodium bicarbonate. However, after they are exposed to light photosynthesis commences in the disks and, with the buildup of oxygen, the disks will either rise to the top of the solution or stand on their side. The length of time that elapses from initial light exposure indicates the rate of O_2 evolution due to photosynthesis. As will be observed, the higher the light intensity, the more rapid the response. This approach may also be used to study the influence of sodium bicarbonate in the infiltration solution, the chlorophyll requirement, and the effect of inhibitors on the process.

Purpose

To demonstrate the leaf disk–sodium bicarbonate infiltration method as used in studies pertaining to the measurement of photosynthetic rates in plant tissue; to measure the effect of several variables on photosynthesis by the leaf disk infiltration method.

Materials

variegated *Coleus* leaves (green-white pattern)

bean leaves (or other thin leaves such as tobacco)

citrate-phosphate buffer, pH 6.8 (see Appendix II)

10^{-3}M $NaHCO_3$ in citrate-phosphate buffer, pH 6.8 (see Appendix II)

10^{-2}M $NaHCO_3$ in citrate-phosphate buffer, pH 6.8 (see Appendix II)

beakers (1000 ml)

cork borer (internal diameter of 1 cm)

infiltration apparatus consisting of several specimen jars, two-hole rubber stopper, glass and rubber tubing, and water aspirator (see Figure 38–1)

petri dishes (5 cm diameter)   ring stand and support
light source (reflector flood lamp,   foot candle meter
   150 watt bulb)   timer

Procedure

A.   Photosynthesis in leaf disks of variegated leaves of *Coleus*

Preparation of leaf disks and infiltration with sodium bicarbonate

All operations should be performed under subdued light conditions.

   Excise several variegated leaves (green-white pattern) from *Coleus* plants and place in a beaker of tap water for one hour. Do *not* expose the detached leaves to bright light. After the soaking period, punch out fifteen disks (about 1 cm in diameter) from the chlorophyll-containing portion of the leaves. Immediately after cutting, place five disks into each of three separate specimen dishes or 125 ml flasks containing 25 ml of 0.01M NaHCO$_3$ in citrate-phosphate buffer (pH 6.8).

   Stopper the containers with a two-hole rubber stopper equipped with glass and rubber tubing attached to a running water aspirator pump, as indicated in Figure 38–1.

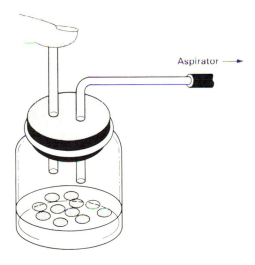

FIGURE 38–1   Leaf disk infiltration.

   Turn on the water and apply suction by holding your index finger over the open glass tube. Then release the vacuum abruptly by removing your finger. Continue the infiltration method until there are no bubbles at the edges of the disks and when they are resting on the bottom of the container. If one or two disks do not respond, replace them with additional disks.

   Pour the infiltrated disks and the suspending solution into a small petri dish (5 cm in diameter) and arrange the disks (use a glass rod) on the bottom so that they are not touching each other. If there are any small bubbles adhering to the disks, tease them off before illumination.

Measurement of photosynthesis

Place the petri dishes containing the disks under a tungsten lamp (turned off). The lamp should be set at a distance above the disks so that they will be exposed to a light intensity of approximately 1500 to 2000 ft.-c.

Turn on the light and record the time in seconds that elapses after turning on the light for each disk to rise to the surface of the liquid or turn on edge. If one or two disks do not respond adequately, they may be discounted.

The time required for the disks to rise or turn on edge is inversely proportional to the photosynthetic rate. Therefore, the photosynthetic rate may be expressed as the reciprocal of time in seconds.

Repeat the same procedures with an equivalent number of disks taken from the white area of *Coleus* leaves. Compare the results with stems of the chlorophyll-containing disks.

Observations:

B. **Effect of light intensity on photosynthetic rate of bean leaf disks**

Follow the same general procedures outlined in Part A. For the experiment, prepare leaf disks from between the main veins of fully expanded detached bean leaves previously soaked in tap water for one hour. (Other thin leaves such as tobacco may be used if desired.)

Using separate sets of disks (five disks per set and about 1 cm in diameter), measure the photosynthetic rate at 250, 500, 1000, and 2000 ft.-c. The different levels of light intensity may be obtained by setting the lamp at various distances above the sample. However, for light intensities greater than 2000 ft.-c, filter them through a glass container of water so that the test solutions do not become heated.

Observations:

C. **Photosynthesis in leaf disks evacuated in distilled water or sodium bicarbonate solution**

Prepare fifteen leaf disks from between the main veins of fully expanded leaves of bean plants (other expanded, thin leaves may be used if desired). Be sure to float the leaves in tap water for one hour prior to cutting. Place five disks in each specimen dish:

Dish 1: citrate-phosphate buffer (pH 6.8)

Dish 2: citrate-phosphate buffer + $NaHCO_3$ (0.001M)

Dish 3: citrate-phosphate buffer + $NaHCO_3$ (0.01M)

Infiltrate the disks in each flask and subject to light (about 2000 ft.-c) according to the procedure outlined previously.

Observations:

**Results/
Conclusions**

Construct a table or graph that illustrates the relative rate of photosynthesis for the disks in each part of the exercise. Make general conclusions concerning the procedures and their results. As a guideline, consider the answers to the following:

1. Why did the green disks rise to the surface of the buffered sodium bicarbonate solution or turn on end when exposed to light?

2.   What aspect of photosynthesis is illustrated by this exercise, and what is the role of chlorophyll?

3.   Is it possible to perform this exercise when the disks are infiltrated with only buffer solutions?

4.   What is the role of the sodium bicarbonate in the performance of this exercise?

5.   What are some of the advantages of the leaf disk infiltration method as compared to the Hill reaction (Exercise 39) for photosynthetic rates of leaf tissue?

Acknowledgment

This exercise was adapted for class use from the work of J. L. Wickliff and R. M. Chasson.

References

Govindjee and J. J. S. van Rensen. 1978. Bicarbonate effects on the electron flow in isolated broken chloroplasts. *Biochem. Biophys. Acta* 505: 183–213.

Gregory, R. P. F. 1977. *Biochemistry of Photosynthesis*, 2nd edition. New York: Wiley.

Haliwell, B. 1981. *Chloroplast Metabolism*. Oxford: Oxford University Press.

Hipkins, M. F. 1984. Photosynthesis. In Wilkins, M. B. (ed.), *Advanced Plant Physiology*. London: Pitman Publishing, Limited, pp. 219–248.

Radmen, R. and G. Cheniae. 1977. Mechanisms of oxygen evolution. In Barber, J. (ed.), *Primary Processes of Photosynthesis, Topics in Photosynthesis,* Vol. 2. Amsterdam: Elsevier, pp. 303–348.

Wickliff, J. L. and R. M. Chasson. 1964. Measurement of photosynthesis in plant tissues using bicarbonate solutions. *Bioscience* 14: 32–33.

Wickliff, J. L. and R. M. Chasson. 1967. Influence of virus infection in photosynthesis. In Sourcebook Committee of the American Phytopathological Society, A. Kelman (Chmn.), *Source Book of Laboratory Exercises in Plant Pathology*. San Francisco and London: W. H. Freeman and Co., pp. 163–165.

EXERCISE 39

The Hill reaction

In 1937, R. Hill, the English biochemist, demonstrated that a chloroplast preparation in the presence of light, water, and a suitable hydrogen acceptor evolved oxygen in the absence of CO_2. The significance of his observation was that the evolution of oxygen is a consequence of the photochemical reactions. Also, the presence of water and the absence of CO_2 supported the idea that water and not CO_2 was the sole source of the oxygen evolved in photosynthesis. Later experiments by others using isotopes supported this assumption. Further, the evolution of oxygen from water made a very strong case for the photolysis (splitting) of water during the light reactions. Today, scientists often refer to the Hill reaction as the photolysis of water in chloroplasts exposed to light with the subsequent evolution of oxygen.

It is interesting to note that Hill in his early experiments used oxyhemoglobin (red) as the electron acceptor, which, when reduced, turned blue. Currently, however, 2,6-dichlorophenol indophenol (DIP) is used as the acceptor for demonstrating the Hill reaction. The DCIP is blue when oxidized, but is colorless when reduced according to the following illustration:

$$\text{DCIP (quinone-blue color)} + H_2O \xrightarrow[\text{chloroplasts}]{\text{light}}$$

$$\text{DCIP–}H_2 \text{ (phenol-colorless)} + \frac{1}{2}O_2 \uparrow$$

In order to assess the extent of the Hill reaction under different conditions, the relative loss of the DCIP (blue color) may be measured colorimetrically.

Purpose

To illustrate the Hill reaction, a light-dependent reaction catalyzed by chloroplasts in which a substance other than nicotinamide adenine dinucleotide phosphate (NADP) is reduced; to determine the effect of atrazine on the Hill reaction.

Materials

10 g fresh spinach leaves

200 ml tricine buffer (0.05M; adjusted to pH 6.5)

0.1 ml atrazine, simazine, or monuron solution (saturated solution)

100 ml 2,6-dichlorophenol-indophenol (1.1×10^{-4}M)

2 ice baths (beakers with chipped ice and NaCl)

water bath (400 ml beaker filled with water at room temperature)

paper towels

scissors

facial tissue

mortar and pestle (chilled)

Buchner funnel

suction flask

centrifuge tubes

clinical centrifuge

ring stand

aluminum foil

5 pipets (10 ml graduated)

2 pipets (1 ml graduated)

8 colorimeter tubes

colorimeter (set at 620 nm)

Procedure

A. Isolation of chloroplasts

Before starting the isolation of the chloroplasts, prepare two baths of chipped ice and NaCl in large beakers. Use these baths to cool *all* reagents, glassware, and the final chloroplast suspension.

Weigh out 10 g of washed, blotted, and cold spinach leaves (use fresh spinach). Shred the leaves with scissors (discard the large veins) and homogenize for two minutes in a chilled mortar with 60 ml of ice-cold 0.05M tricine buffer adjusted to pH 6.5.

Line a Buchner funnel with three buffer-wetted facial tissues and filter the homogenate under vacuum. Collect the filtrate and discard the residue.

Centrifuge the filtrate for four to five minutes at full speed in a clinical centrifuge. After discarding the supernatant, wash the chloroplast pellet with 20 ml ice-cold buffer and recentrifuge. Discard the supernatant, wash the pellet again with 20 ml buffer, and after centrifugation discard the supernatant and resuspend the pellet in 8 ml of 0.05M tricine buffer. While stirring the suspension, make sure the container is immersed in an ice bath.

B. The Hill reaction

Place a 400 ml beaker filled with tap water (20°C) on a ring stand in line with and approximately 6 in. away from a 100 watt lamp.

Under subdued laboratory light, pipet 5 ml of 2,6-dichlorphenol-indophenol (1.1×10^{-4}M) and 5 ml of tricine buffer into each of three colorimeter or spectrophotometer tubes.

Wrap one of the tubes with aluminum foil so that no light can enter. Place all tubes into the 400 ml water bath for two minutes. Then pipet 0.5 ml of the chloroplast suspension into each tube and mix by inversion. Place all tubes back into the water bath, turn on the light, and illuminate the tubes for five minutes.

After five minutes, measure the absorbancy (use a colorimeter set at 620 nm) of the mixture in *one* of the tubes exposed to light and the one wrapped in aluminum foil (control) as compared to a water blank. After reading this latter tube, add a few milligrams of ascorbic acid and note the change in the color of the reaction mixture. At the end of ten minutes (total time from the beginning), take a reading of the remaining tube against a water blank.

For a correct measure of the Hill reaction, subtract the optical density readings obtained for each light-exposed tube from that of the control (aluminum foil-covered).

C.   Effect of atrazine on the Hill reaction

Pipet the necessary components into appropriately labeled colorimeter tubes as outlined in Table 39–1. Before adding the chloroplast suspension, place all tubes into the water bath (25°C) for two minutes. Then pipet the chloroplast suspension into the tubes, mix by inversion, return to the water bath, and illuminate the tubes for ten minutes.

After ten minutes, measure the absorbency of each mixture against a water blank (colorimeter set at 620 nm). Record the readings for each mixture and the adjusted values (from the control) in the space provided in Table 39–1.

TABLE 39–1   Effect of atrazine on the Hill reaction

Components*	Tube Number				
	1	2	3[†]	4[††]	5
0.05M tricine buffer (pH 6.5)	5.0	5.0	5.0	5.0	4.9
2,6-DCIP (1.1×10^{-4}M)	None	5.0	5.0	5.0	5.0
Distilled water	5.0	None	None	None	None
Atrazine (saturated solution)	None	None	None	None	0.1
Chloroplast suspension	0.5	0.5	0.5	(Boiled) 0.5	0.5
Light duration (minutes)	10	10	Dark	10	10

Colorimeter readings

Adjusted readings (from the control)

* Number given for the components represents the amount of each in milliliters.
† Tube 3 should be covered with aluminum foil.
†† Inactivate a portion of the chloroplast suspension by boiling ten minutes, allowing to cool, and adjusting back to volume with buffer before adding to tube 4.

D. Time course study of the Hill reaction

If time permits, design and perform an experiment that illustrates the extent of the Hill reaction with time. Remember to use suitable controls, appropriate reaction components, and the basic techniques outlined previously.

In the space provided, present your experimental design and/or the results of your experiment.

Experimental design and results:

Results/ Conclusions

Present the results obtained for each part of this exercise and, if a time course study was performed, present a graph in which the rate of the Hill reaction is plotted as the ordinate against time on the abscissa. (Use initial OD values, corrected values from the control, or the amount of dye reduced.)

Interpret your results and by a conventional equation illustrate the details of the Hill reaction. Consider the role of the chloroplasts, light, water, tricine buffer, and 2,6-dichlorophenol-indophenol. What is the effect of atrazine on the Hill reaction? Present any current explanations concerning the effect of atrazine or other triazines on photosynthesis. Do other herbicides affect the Hill reaction?

Acknowledgment

This exercise was adapted for class use by C. J. Pollard, Michigan State University, East Lansing.

References

Arnon, D. I. 1971. The light reactions of photosynthesis. *Proc. Nat. Acad. Sci. U.S.* 68: 2883–2892.

Clayton, R. K. 1980. *Photosynthesis: Physical Mechanism and Chemical Patterns.* Cambridge: Cambridge University Press.

Hill, R. 1937. Oxygen evolved by isolated chloroplasts. *Nature* 139: 881.

Hill, R. and R. Scarisbrick. 1940. Production of oxygen by isolated chloroplasts. *Nature* 146: 61–62.

Holt, A. S., R. E. Smith, and C. S. French. 1951. Dye reduction by illuminated chloroplasts. *Plant Physiol.* 26: 164–175.

Nicholls, D. G. 1981. *Bioenergetics: An Introduction to the Chemiosmotic Theory.* New York: Academic Press.

Radmer, R. and G. Cheniae. 1977. Mechanisms of oxygen evolution. In Barber, J. (ed.), *Primary Processes of Photosynthesis: Topics in Photosynthesis,* Vol. 2. Amsterdam: Elsevier, pp. 303–348.

(Also see the references following Exercises 37 and 38.)

EXERCISE 40

Pentose-5-phosphate isomerase

Introduction

Ribose-5-phosphate (pentose-5-phosphate) and ribulose-5-phosphate are important compounds in photosynthesis (carbon dioxide fixation according to the Calvin-Benson scheme) and in the reaction of the hexose monophosphate shunt (pentose phosphate cycle). As illustrated by the Calvin-Benson scheme of carbon dioxide fixation, ribose-5-phosphate is converted to ribulose-5-phosphate, which in turn is the immediate precursor to ribulose-1,5-bisphosphate (RuBP). This latter compound is important in the fixation of carbon dioxide. Each molecule of ribulose 1,5-bisphosphate fixes one molecule of carbon dioxide with the addition of water and the ultimate formation of two molecules of 3-phosphoglyceric acid.

The conversion of ribose-5-phosphate to ribulose-5-phosphate is mediated by the enzyme ribose-5-phosphate isomerase. This enzyme is found in high quantities in chloroplasts as would be expected for its activity in the formation of ribulose-5-phosphate.

The assay for the formation of ribulose-5-phosphate is determined by the development of red color after the addition of resorcinol–hydrochloric acid reagent to the incubation solution and reaction with the ketose (ribulose-5-phosphate). In order to provide a measure of the relative extent of each reaction, the relative amounts of colored product may be determined by taking absorbance (optical density) readings at 420 nm with a spectrophotometer or colorimeter.

Purpose

To prepare a crude extract from spinach leaves containing pentose-5-phosphate isomerase and to study the enzymatic conversion of a phosphorylated aldose (ribose-5-phosphate) to a phosphorylated ketose (ribulose-5-phosphate).

Materials

50 g spinach leaves

500 ml tris buffer (0.1M; adjusted to pH 7.0 with HCl)

5 ml ribose-5-phosphate solution (10 mg/5 ml water; w/v; the sodium salt works best)

20 ml resorcinol–hydrochloric acid reagent (see Appendix II)

blender

facial tissue

centrifuge tubes

centrifuge

5 test tubes

water bath

colorimeter or spectrophotometer

Klett colorimeter (420 filter) or spectrophotometer (set at 420 nm)

Procedure

A. Enzyme preparation

The extract may be prepared before the class meeting with an aliquot distributed to each member or group.

Homogenize 50 g of spinach leaves with 200 ml of cold, 0.1M tris buffer (pH 7.0) in a blender for two minutes. Filter the homogenate through two layers of buffer-wetted facial tissue by gravity. Discard the residue and centrifuge the filtrate at 15,000 × g for twenty minutes. Use the resulting supernatant as a source of the enzyme, but dilute with tris buffer ten times before use. (It may be wise to use several different dilutions.)

If the enzyme preparation is not to be used immediately, it may be frozen and stored for a short duration.

B. Pentose-5-phosphate isomerase assay

Label five clean and dry test tubes 1 to 5, and to each tube add the various components according to the following sequence (as a final check see Table 40–1):

1. Pipet 0.6 ml of 0.1M tris buffer into test tubes 1, 2, and 3.

2. Pipet 0.8 ml of 0.1M tris buffer into test tubes 4 and 5.

3. Pipet 0.2 ml of the diluted enzyme solution into tubes 1, 2, 3, and 5.

4. Heat tube 3 *only* in a boiling water bath for two minutes and then cool the tube rapidly in cold water. If necessary, add sufficient buffer to replace any liquid which evaporated from the tube.

5. Start the reaction by adding 0.2 ml of ribose-5-phosphate solution into tubes 1, 2, 3, and 4.

6. After adding the substrate, mix the contents of each tube and incubate at room temperature for fifteen minutes.

TABLE 40–1   Pentose-5-phosphate isomerase reaction mixture*

	Tube Number and Amount Added (ml)				
Stock Solution	1	2	3	4	5
Tris buffer (0.1M, pH 7.0)	0.6	0.6	0.6	0.8	0.8
Enzyme preparation	0.2	0.2	0.2 (Boiled)	—	0.2
Ribose-5-phosphate	0.2	0.2	0.2	0.2	—

* Incubate all mixtures for 15 minutes.

Stop the reaction at the end of incubation time by adding 4 ml of resorcinol-hydrochloric acid reagent from a burette into each of the test tubes. Shake the tubes and heat them in a water bath while maintaining the temperature between 75 and 80°C for *exactly* ten minutes. Then cool the tubes in cold water and measure the absorbance of each mixture with a Klett colorimeter (420 filter) or spectrophotometer (set at 420 nm). Consult the instructor for the proper use and standardization of the instruments and remember to use a water blank.

Results/ Conclusions

Record the readings (Klett or optical density units) in Table 40–2.

TABLE 40–2   Ribulose-5-phosphate determination

	Tube Number				
Treatment or Measurement	1	2	3	4	5
Resorcinol-HCl (ml)	4	4	4	4	4
Reaction time (min)	10	10	10	10	10
Klett units (420 filter) or					
Optical density (420 nm)					
Adjusted value (enzyme activity)			—	—	—

Since the slow conversion of ribose-5-phosphate to ribulose-5-phosphate takes place in the absence of enzyme, calculate the enzyme activity by taking the difference in the readings of each experimental mixture containing boiled enzyme (tube 3). Record the adjusted values for tubes 1 and 2 in Table 40–2.

In concluding remarks, consider the procedures involved in the enzyme preparation. For example, explain why cold tris buffer (pH 7.0) was used as the extracting medium. Is the pentose-5-phosphate isomerase the only enzyme present in the final extract? Why were mixtures 4 and 5 included in the enzyme assay experiment?

Present a general equation illustrating the details of the enzymatic conversion of ribose-5-phosphate to ribulose-5-phosphate. Is light required for the reaction or are other factors involved in carbon dioxide fixation? What is the importance of this reaction in respiration and in photosynthesis?

Illustrate the reaction details involved in the detection of ribulose-5-phosphate when resorcinol-hydrochloric acid reagent is used. Is the reagent specific for the detection of ribulose-5-phosphate or for a reactive chemical group? Explain.

What further procedure might be used to ascertain conclusively that ribulose-5-phosphate was produced in the reaction mixtures containing the enzyme preparation?

Acknowledgment   This exercise was designed for class use by C. J. Pollard, Michigan State University, East Lansing.

References   Axelrod, B. and R. Jang. 1954. Purification and properties of phosphoribo-isomerase from alfalfa. *J. Biol. Chem.* 209: 847–855.

Bassham, J. A. 1971. Photosynthetic carbon metabolism. *Proc. Nat. Acad. Sci. U.S.* 68: 2877–2882.

Bassham, J. A. and M. Calvin. 1957. *The Path of Carbon in Photosynthesis.* Englewood Cliffs, N. J.: Prentice–Hall, Inc.

Gibbs, M. and E. Latzko (eds.). 1979. Photosynthetic carbon metabolism and related processes. Phytosynthesis. II. In *Encyclopedia of Plant Physiology*, New Series, Vol. 6. New York: Springer-Verlag.

Haliwell, B. 1981. *Chloroplast Metabolism*. Oxford: Oxford University Press.

Siegelman, H. W. and G. Hind (eds.). 1978. *Photosynthetic Carbon Assimilation.* New York: Plenum Press.

Photosynthetic rates of intact plant leaves

Introduction

Photosynthetic rates of intact leaves may be determined by means of an open infrared carbon dioxide gas exchange system. If such an instrument is available, be sure to become familiar with its operation before use. Essentially, the system consists of a chamber in which an intact leaf of a small plant is placed. The leaf sealed in the chamber may be exposed to high-intensity light at constant temperature. A gas mixture containing carbon dioxide, oxygen, and nitrogen in known amounts is passed through the chamber. Depending upon the system, the concentration of CO_2 may be determined before and after passage through the chamber. The amount of CO_2, however, will be analyzed by passage through an infrared analyzer according to procedures established by the instructor.

The procedures used should be directed toward the measurement of CO_2 before and after passage of the gas mixture into and out of the leaf chamber. In this way, one can assess the amount of CO_2 taken up by the leaf and hence the rate of photosynthesis.

Purpose

To determine the photosynthetic rates of intact leaves with an open infrared carbon dioxide exchange system.

Materials

open infrared gas exchange system (see Figure 41–1 or special instructions about the one available in your laboratory), equipped with:

- tank of oxygen, carbon dioxide, and nitrogen
- gas mixer
- reference flow meter
- H_2O scrubber (dehydrite)
- CO_2 scrubber (ascarite)
- pre-flow meter
- O_2 sensors

- infrared gas analyzer
- post-flow meter
- humidity sensor
- high-intensity light source
- leaf chamber
- water bath
- temperature controller
- high-intensity lamp

several of the following small potted plants with fully expanded leaves, approximately three to four weeks old and representing C_3 and C_4 metabolism:

—monocotyledons (C_3) such as wheat (*Triticum aestivum* L.) or oat (*Avena sativa* L.)

—dicotyledons (C_3) such as Mung bean (*Vigna radiata* L.) or squash (*Cucurbita moschata* L.)

—monocotyledons (C_4) such as corn (*Zea mays* L.) or sorghum (*Sorghum vulgare* L.)

—dicotyledons (C_4) such as kochia (*Kochia scoparis* L.)

Procedure

An open infrared CO_2 gas exchange system is illustrated in Figure 41–1. Although the system may vary from one laboratory to another, essentially the same components are used for the determination of photosynthetic rates. You should learn the basic components and operation of the available equipment before use.

After becoming familiar with the system, select a fully expanded leaf on a known C_3 or C_4 plant. Determine the surface area (dm^2), and place it into the leaf chamber according to the direction of the instructor. The chamber with the intact leaf should be exposed to high light levels with the temperature in the chamber held constant at 25°C (C_3 plants) or 30°C (C_4 plants).

Once the chamber is sealed with the leaf in place and exposed to high-intensity light, close the valve at points 2 and 4 (Figure 41–1) and open the valve at point 3. Then start a gas mixture containing 300 ppm CO_2, 21% O_2, and the remainder N_2 through the system. The mixture will pass through the gas mixer, flow meters, O_2 sensors, into the infrared gas analyzer, and to the atmosphere. Take several readings from the meters on the O_2 sensors and infrared gas analyzer to determine pre-chamber concentrations of CO_2 and O_2.

After ensuring that the system is operating, open the valves at points 2 and 4 and close the valve at point 3. The gas mixture will then pass into the leaf chamber and out through the humidity sensor, O_2 sensors, dehydrite chamber, post-flow meter, infrared gas analyzer, and then to the atmosphere.

The flow rate of the mixture may have to be adjusted so that it is fast enough to avoid depletion of the CO_2 in the chamber but not so fast as to inhibit photosynthesis due to stomatal closure. If more than 15 ppm CO_2 is taken out of the mixture in the leaf chamber (difference between the pre- and post-chamber readings), the flow rate is too slow. If the difference between pre- and post-chamber readings is negligible, photosynthesis is being inhibited and the flow rate is too fast. Since the flow rate necessary for photosynthesis will vary with the plants used, it must be monitored carefully. Once a suitable flow rate has been established, determine pre-chamber readings again followed by measurements of CO_2 exchange in the chamber.

Following the above procedures, measure the photosynthetic rates (gas exchange) for several leaves on the same (C_3) and then on a different plant (C_4). Be sure to obtain several readings for comparison. Also, the instructor may have you

190

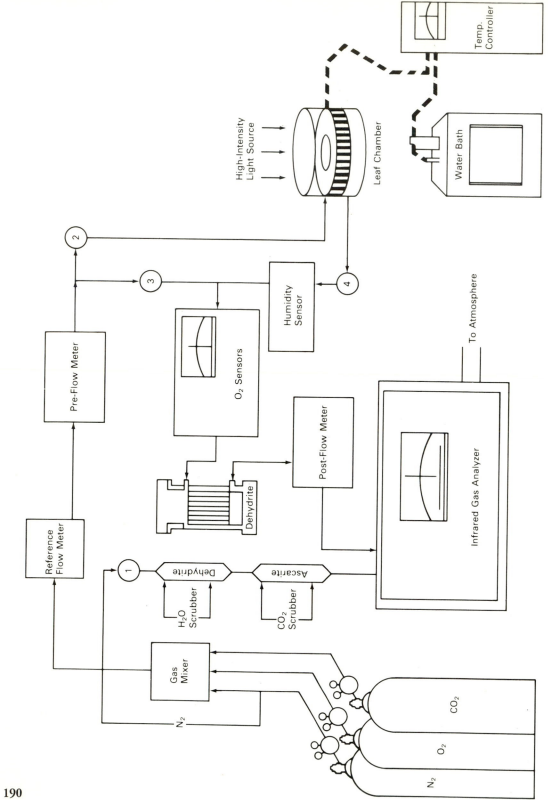

FIGURE 41-1   Open infrared CO_2 gas exchange system.

perform additional experiments. For example, the effect of temperature and different CO_2 and/or O_2 concentrations may be studied.

The difference in CO_2 concentration (pre- and post-chamber) of the gas mixtures is used to calculate the photosynthetic rate (mg CO_2 taken up per dm^2 per hr.) according to the following equation:

$$Pn = \frac{CF \times ppm\ CO_2}{T\ (^\circ K)} \times \frac{1}{(dm^2)} \times \frac{J}{T\ (^\circ K)} \times \frac{T\ (^\circ K)}{T^1} \times CF^*$$

where:

Pn = photosynthetic rate

CF = amount of CO_2 (mg/liter) in air at different CO_2 concentrations (μl l^{-1}) and temperatures (see Table 41–1.)

ppm CO_2 = difference between CO_2 of pre- and post-chamber gas mixture

$T\ (^\circ K)$ = 273 plus temperature ($^\circ C$) of chamber

$1/(dm^2)$ = leaf area

J = flow rate ($l\ hr.^{-1}$)

T^1 = temperature in flow meter (usually same as ambient or room temperature)

$$CF^* = \frac{molecular\ weight\ of\ CO_2 \times 273}{22.414}$$

Let us now consider a hypothetical situation to calculate the photosynthetic rate of a leaf. An intact leaf (leaf area = 0.1 dm^2) of a C_4 plant absorbed sufficient CO_2 to cause a difference in CO_2 concentration between the pre- and post-chamber gas mixture of 36 ppm. The temperature of the leaf chamber was 30°C. The temperature in the flow meter was the same as ambient temperature (20°C). The flow rate was 59.1 l/hr. Using Table 41–1 and the data presented, the following equation would be set up accordingly:

$$Pn = \frac{0.534 \times 36}{303} \times \frac{1}{0.1} \times \frac{59.1}{303} \times \frac{303}{293} \times 535.91$$

$$= 0.063 \times 10 \times 0.195 \times 1.017 \times 535.91$$

$$= 6.69\ mg\ CO_2/dm^2/hr.$$

Results/ Conclusions

Record your results, highlighting differences in photosynthetic rates between representative C_3 and C_4 plants. You should consider the reliability of the apparatus used, any problems that might be inherent in the technique, and the reasons for differences between C_3 and C_4 plants with respect to CO_2 uptake.

Any other data for additional experiments should be recorded. In addition, consider the effects of temperature and increased CO_2 concentration or oxygen on photosynthetic rates. Also consider why the stomates of leaves in the chamber close if the flow rates are too excessive.

TABLE 41-1   Amount of CO_2 (mg/liter) in air of different CO_2 concentrations ($\mu l\ l^{-1}$) at different temperatures

CO_2 ($\mu l\ l^{-1}$)	mg $CO_2\ l^{-1}$			
	15°C	20°C	25°C	30°C
200	0.375	0.368	0.362	0.356
220	0.412	0.405	0.398	0.392
240	0.450	0.442	0.435	0.427
260	0.487	0.479	0.471	0.463
280	0.525	0.516	0.507	0.499
300	0.562	0.553	0.543	0.534
320	0.600	0.589	0.580	0.570
340	0.637	0.626	0.616	0.606
360	0.675	0.663	0.652	0.641
380	0.712	0.700	0.688	0.677
400	0.750	0.737	0.724	0.712
420	0.787	0.774	0.761	0.748
440	0.825	0.810	0.797	0.784
460	0.862	0.847	0.833	0.819
480	0.900	0.884	0.869	0.855
500	0.937	0.921	0.906	0.891

Acknowledgment   This exercise was adapted for class use by R. N. Arteca, The Pennsylvania State University, University Park.

References   Arteca, R. and C. N. Dong. 1982. Increased photosynthetic rates following gibberellic acid treatments to the roots of tomato plants. *Photosynthesis Res.* 2: 243–249.

Black, C. C. 1973. Photosynthetic carbon fixation in relation to net CO_2 uptake. *Ann. Rev. Plant Physiol.* 24: 253–286.

Edwards, G. and D. A. Walker. 1983. *C-3, C-4 Mechanisms and Cellular and Environmental Regulation of Photosynthesis.* Oxford: Blackwell.

Haliwell, B. 1981. *Chloroplast Metabolism.* Oxford: Oxford University Press.

Hipkins, M. F. 1984. Photosynthesis. In Wilkins, M. B. (ed.), *Advanced Plant Physiol.* London: Pitman Publishing, Limited, pp. 219–248.

Lorimer, G. H. 1981. The cargoxylation and oxygenation of ribulose-1,5-bisphosphate: The primary events in photosynthesis and photorespiration. *Ann. Rev. Plant Physiol.* 32: 349–383.

Sestak, Z., J. Catsky, and P. G. Jarvis (eds.). 1971. *Plant Photosynthetic Production, Manual of Methods.* The Hague: JUNK.

EXERCISE 42

The effect of chemicals on lettuce seed germination

Introduction

There are numerous chemical promotors of germination, the most notable of which include potassium nitrate, thiourea, ethylene, gibberellins, and kinetin. Thiourea, gibberellins, and kinetin are interesting in that in certain instances they substitute for the light requirements in light-sensitive seeds. Even though true substitution is disputed, they do show promotion of germination of seeds in darkness.

Many compounds inhibit germination due to their toxicity to essential life processes. The types of inhibitors important in regulating germination fall into the category of compounds naturally occurring in seeds. Some of these compounds are often the cause of dormancy and possibly act to block some essential germination processes. Natural germination inhibitors do not reduce the viability of the seed, however, and can be overcome under the right conditions (leaching out of seeds, decreased biosynthesis or chemical structural changes).

Natural inhibitors may be found in different structures, including the pulp or juice of fruit, the seed coat, the endosperm, the embryo, and so on. Some of the natural germination inhibitors include abscisic acid, coumarin, parascorbic acid, ammonia, phthalids, and ferulic acid.

Of the synthetic compounds that retard germination, the commercial herbicides are potent inhibitors. The herbicide 2,4-dichlorophenoxyacetic acid (2,4-D) is an example of such a herbicide.

A very simple way to test compounds for activity either as a promotor or inhibitor of germination is to sow seeds on filter paper wetted with a solution of the compound, followed by incubation. These procedures are outlined in the following exercise.

Purpose

To determine the effect of various chemicals on lettuce seed germination.

Materials

450 lettuce seeds (*Lactuca sativa* L. 'Grand Rapids')

9 petri dishes lined with filter paper disks

7 pipets (graduated, 5 ml)

aqueous solution of the following:
- coumarin (5 mg/100 ml)
- thiourea (125 mg/100 ml)
- urea (125 mg/100 ml)
- actinomycin D (5 mg/100 ml)
- 2,4-dichlorophenoxyacetic acid (7 mg/l)
- 2,4-dinitrophenol (125 mg/100 ml)

2 incubators (set at 20°C and 32°C)

Procedure

Number the petri dishes 1 to 9 and place a filter paper disk on the bottom of each. Then place fifty lettuce seeds (*Lactuca sativa* L. 'Grand Rapids') on each dish and pipet on the seed 4 ml of the following solutions:

Dish 1: distilled water

Dish 2: coumarin (5 mg/100 ml)

Dish 3: thiourea (125 mg/100 ml)

Dish 4: urea (125 mg/100 ml)

Dish 5: actinomycin D (5 mg/100 ml)

Dish 6: 2,4-dichlorophenoxyacetic acid (7 mg/l)

Dish 7: 2,4-dinitrophenol (125 mg/100 l)

Dish 8: distilled water

Dish 9: thiourea (125 mg/100 ml)

Allow the seeds to imbibe the solutions for thirty minutes while irradiating with red light (a red fluorescent lamp covered with two layers of red cellophane may be used).

Place dishes 1 through 7 in a 20°C incubator. Place dishes 8 and 9 in a 32°C incubator. Allow the seeds to germinate for seventeen to twenty-four hours and count the number of seeds germinated in each dish.

Results/ Conclusions

Using the data obtained, fill in Table 42–1 and present as part of the results.

In concluding remarks, evaluate and interpret the data obtained from this experiment. If statistical evaluations of the results are desired, use the total class results.

TABLE 42–1   Lettuce seed germination

Dish	No. of Seeds Planted	No. of Seeds Germinated	Percent Germination	Percent Inhibition
1	50			
2	50			
3	50			
4	50			
5	50			
6	50			
7	50			
8	50			
9	50			

Acknowledgment   This exercise was designed for class use by C. J. Pollard, Michigan State University, East Lansing.

References   Ashton, F. M. 1976. Mobilization of storage proteins of seeds. *Ann. Rev. Plant Physiol.* 27: 95–117.

Berrie, A. M. 1984. Germination and dormancy. In Wilkin, M. B. (ed.), *Advanced Plant Physiology*. London: Pitman Publishing, Limited, pp. 440–468.

Bewley, J. D. and M. Black. 1982. *Physiology and Biochemistry of Seeds in Relation to Germination*, Vol. 2. Berlin: Springer-Verlag.

Dure, L. S. 1975. Seed formation. *Ann. Rev. Plant Physiol.* 26: 259–278.

Khan, A. A. (ed.). 1982. *The Physiology and Biochemistry of Seed Development, Dormancy and Germination*. Amsterdam: Elsevier Biomedical Press.

Koller, D. 1959. Germination. *Sci. Amer.* 200(4): 75–84.

Mayer, A. M. and A. Poljakoff-Mayber. 1980. *The Germination of Seeds,* 3rd edition. Oxford: Pergamon Press.

Talorson, R. B. and S. B. Hendricks. 1977. Dormancy in seeds. *Ann. Rev. Plant Physiol.* 28: 331–354.

Thompson, R. C. and W. F. Kosar. 1938. The germination of lettuce seed stimulated by chemical treatment. *Science* 87: 218–219.

The influence of kinetin and light on lettuce seed germination

Borthwick, Hendricks, and colleagues (see the references that follow) originally demonstrated that red light (660 nm) promoted the germination of seeds of lettuce (*Lactuca sativa* L. 'Grand Rapids'). They further showed that far-red light (735 nm) following the red light exposure inhibited the process and that the phytochrome pigment system was involved in the light reception process.

It is now well established that phytochrome is a pigment involved in the reception of photoperiodic stimuli controlling flowering as well as lettuce seed germination and other morphogenetic phenomena. Simply stated, phytochrome exists in two forms, the phytochrome red absorbing form (Pr) and the phytochrome far-red absorbing form (Pfr). The Pfr form is likely the physiologically active form, although the two forms are interconvertible. In addition, the Pfr form appears to be slowly converted to the Pr form in darkness or when exposed to far-red light. The Pfr form is also believed to decay to an unknown inactive form. This decay process refers only to a loss of photoreversibility and not to actual destruction.

There seems to be a number of short-lived forms or intermediates between Pr and Pfr. Thus, when one form converts to the other form, the conversion occurs through a number of transient intermediates. Further explanations of the chemistry of phytochrome is given in a number of texts and articles on this subject.

When light-sensitive lettuce seeds that are imbibing water are treated with red or white light, the Pfr form accumulates and is translated into a germination response. However, if the red light treatment is followed by far-red light, the Pfr form is converted to the inactive Pr form and germination does not take place. Cytokinins, however, will stimulate germination in darkness and to some extent promote germination, even when imbibing seeds are exposed to far-red light. This phytohormone-induced response was demonstrated by Miller (see the references). Although the mechanism through which the cytokinins operate is still unknown, they probably operate on growth responses removed from the phytochrome system at some point in the sequence of reactions required for germination.

Purpose

To observe the effect of kinetin, red light, and far-red light on lettuce seed germination in the dark.

Materials

350 lettuce seeds (*Lactuca sativa* L. 'Grand Rapids')	darkroom
2 kinetin solutions (2 mg/l and 10 mg/l; w/v)	2 small cardboard boxes (any suitable light-tight container to hold and transport the petri dishes)
7 petri dishes, each containing three disks (9 cm) of Whatman No. 1 filter paper	green safelight (see Appendix II)
aluminum foil (or black sateen cloth)	red light source (see Appendix II)
	far-red light source (see Appendix II)

Procedure

A. Plant material

To ensure relatively uniform germination, incubate the dry lettuce seeds (*Lactuca sativa* L. 'Grand Rapids') at 35 to 37°C for three days prior to the experiment.

B. Chemical and light treatment of the seeds

Label seven petri dishes, each containing pads of filter paper (three disks of Whatman No. 1, 9 cm). Then add 5 ml of water or kinetin solution to each dish according to Table 43–1. Count out and sprinkle fifty Grand Rapids lettuce seeds into each dish. Cover the dishes and wrap them individually with aluminum foil

TABLE 43–1   Effect of kinetin, red light, and far-red light on lettuce seed germination*

Dish	Test Solution (mg/l)	Light Treatment	No. of Seeds Germinated	Percent Germination
1	Water	None		
2	Kinetin (2)	None		
3	Kinetin (10)	None		
4	Water	Red (8 min.)		
5	Water	Red (8 min.) followed by far-red (8 min.)		
6	Water	Far-red (8 min.)		
7	Kinetin (10)	Far-red (8 min.)		

* Fifty seeds sprinkled into each dish.

or black sateen cloth. Remember that for successful results, the imbibing seeds are not to be exposed to any light unless otherwise indicated. Also, on the outside of each petri dish wrapping, place small strips of masking tape (according to your own coding system) so that each dish may be distinguished in a darkroom under a green safelight.

Petri dishes 1, 2, and 3 may be placed in a container where they are to remain in the dark for the full duration of the experiment (forty-eight hours). Petri dishes 4 through 7 are to be placed in another container and stacked from bottom to top in the following order: 7, 6, 5, and 4.

Store both containers in the dark; from now on all manipulations should be performed in the dark. Even though a green safelight may be used for the following operations, it should be realized that even dim green light will enhance the kinetin effect on lettuce seed germination (Miller 1958).

At the end of sixteen hours' imbibing time, remove petri dishes 4 and 5 from the box and in the dark unwrap and uncover each dish. Then expose the seeds to eight minutes of red light.

After the red light exposure, cover and rewrap petri dish 4 only. Now in the same manner expose the seeds in petri dishes 5, 6, and 7 to far-red light for exactly eight minutes. Return all petri dishes to total darkness for the remaining hours of experimental time.

At the end of forty-eight hours from the beginning of the experiment, unwrap all the petri dishes and in the laboratory count the number of germinated seeds in each petri dish. Record the number and percentage of germinated seeds for each treatment in Table 43–1. Since kinetin at the concentrations used inhibits radicle elongation, a seed may be regarded as germinated if any part of the embryo protrudes through the seed coat.

Results/Conclusions

Present Table 43–1 and a suitable bar graph relating the percent germination to the various kinetin and light treatments in the results section of your report. Pooled class results will give more information for statistical purposes.

Then use the illustrated data for a discussion pertaining to the effects of kinetin, red light, and far-red light on lettuce seed germination in the dark. Consider the pigment system involved in the light effects and any possible mechanisms involving the effect of kinetin on lettuce seed germination.

Acknowledgment

The exercise outlined here and based on the effect of kinetin and light on lettuce seed germination was adapted from the work of C. O. Miller. Work specifically relating the effects of light on lettuce seed germination is indicated in the reference section (for example, see Borthwick, Flint, and Hendricks).

References

Berrie, A. M. 1984. Germination and dormancy. In Wilkins, W. B. (ed.), *Advanced Plant Physiology*. London: Pitman Publishing, Limited, pp. 440–458.

Bewley, J. D. and M. Black. 1978. *Physiology and Biochemistry of Seeds in Relation to Germination*, Vol. 1. Berlin: Springer-Verlag.

Borthwick, H. A., S. B. Hendricks, M. W. Parker, E. H. Toole, and V. K. Toole. 1952. A reversible photoreaction controlling seed germination. *Proc. Nat. Acad. Sci. U.S.* 38: 662–666.

Borthwick, H. A., S. B. Hendricks, E. H. Toole, and V. K. Toole. 1954. Action of light on lettuce seed germination. *Bot. Gaz.* 115: 205–225.

Flint, L. H. and E. D. McAlister. 1937. Wavelengths of radiation in the visible spectrum promoting the germination of light sensitive lettuce seed. *Smithsonian Inst. Publs. Misc.,* Coll. 96: 1–8.

Hendricks, S. P. 1964. Photochemical aspects of plant photoperiodicity. In Giese, A. C. (ed.), *Photophysiology.* New York: Academic Press.

Miller, C. O. 1956. Similarity of some kinetin and red-light effects. *Plant Physiol.* 31: 318–319.

Skinner, C. G., J. R. Claybrook, F. D. Talbert, and W. Shive. 1956. Stimulation of seed germination by 6-(substituted) thiopurines. *Arch. Biochem. Biophys.* 65: 567–569.

Taylorson, R. B. and S. B. Hendricks. 1977. Dormancy in seeds. *Ann. Rev. Plant Physiol.* 28: 331–354.

Toole, E. H., S. B. Hendricks, H. A. Borthwick, and V. K. Toole. 1956. Physiology of seed germination. *Ann. Rev. Plant Physiol.* 7: 299–324.

EXERCISE 44

Photoperiodism in cocklebur

Introduction

Cocklebur (*Xanthium*), a short-day flowering plant, has a critical day length of approximately fifteen-and-a-half hours and will flower if exposed to photoperiods not exceeding the critical day length.

Actually, flowering is more a response to the dark period than to the photoperiod. Short-day plants such as *Xanthium* will flower when a critical dark period is exceeded, while long-day plants flower when the duration of the dark period is less than a critical period.

While the length of the dark period determines the actual initiation of floral primordia, the length of the light period influences the number of floral primordia initiated. In addition, the number of cycles needed to induce flowering differs widely among different plant species. For example, *Xanthium pennsylvanicum* requires only one photoinductive cycle to initiate floral primordia. In contrast, *Salvia occidentalis*, a short-day plant, requires at least seventeen photoinductive cycles to flower. However, once a plant has received the minimum number of photoinductive cycles, it will flower even if it is returned to noninductive cycles. An implication from these observations is that some factor involved in the flowering response is accumulated during the inductive cycle. In some plants, enough factor is accumulated after only one cycle or several to promote flowering. Further, existing evidence shows that the leaves are the organs of perception in the flowering response to photoinductive cycles.

Xanthium plants are ideal for studying photoperiodism and flowering since the flowering response may be detected by examining the changes in buds soon after photoinduction. These changes are due to the conversion of vegetative to floral primordia and are illustrated within the following exercise.

Purpose

To study the photoinduction of flowering in day-length sensitive *Xanthium* plants.

Materials

32 potted cocklebur plants (*Xanthium pennsylvanicum*), sixty days old

greenhouse bench under constant light conditions (if available, a

growth chamber with controlled
lighting)
darkroom, suitable dark box covered with black sateen cloth, or

individual light-tight cabinets
for storing plants according to
treatment

Procedure

A. Cocklebur seed germination and plant growth

Seeds from day-length sensitive cocklebur (*Xanthium pennsylvanicum*) may be collected in the northern latitudes of the United States during the fall of the year. However, if cocklebur is not available, seeds of other light-sensitive plants (*Ipomoea nil* 'Violet,' for example) may be purchased from a local dealer and used according to the general scheme of this experiment. The specific light and dark treatments outlined for *Xanthium* should be changed according to the plants being used.

Before planting, give the cockleburs a cold treatment at 4°C for seventy-two hours. Then soak them in distilled water for twenty-four hours. After the pretreatments, plant the seeds (there is no need to remove the burs) in flats of sand to a depth of about one inch.

When the seedlings have grown to a height of about two inches above the sand, pot them in soil (one seedling/pot) and maintain them under continuous fluorescent light for sixty days.

B. Critical day-length for photoinduction

Expose ten potted sixty-day-old *Xanthium* plants (two/treatment) to the light and dark conditions outlined in Table 44–1. The light and dark intervals should be strictly adhered to for five cycles (days) after which time place all plants under continuous fluorescent light. The number of plants (two/treatment) used for an individual experiment will be sufficient for interpretation only if the members of the class pool their results.

TABLE 44–1 Critical day-length for photoinduction

Pot Number	Light (hr/cycle)	Dark (hr/cycle)	Number of Cycles (days)	Class Results *	
				1	2
B-1	10	14	5		
B-2	15	9	5		
B-3	15½	8½	5		
B-4	16	8	5		
B-5	20	4	5		

* Use an average value for the class results as determined by Methods 1 and 2 outlined in the Results section.

C. Number of inductive cycles necessary for flowering

Expose twelve sixty-day-old *Xanthium* plants to the light and dark cycles outlined in Table 44–2. After exposing the plants to the experimental conditions for the prescribed number of cycles, return them to long-day conditions under the lights.

TABLE 44–2 Number of inductive cycles necessary for flowering

Pot Number	Light (hr/cycle)	Dark (hr/cycle)	Number of Cycles (days)	Class Results*	
				1	2
C-1	24	0	5		
C-2	8	16	1		
C-3	8	16	2		
C-4	8	16	3		
C-5	8	16	4		
C-6	8	16	5		

* Use an average value for the class results as determined by Methods 1 and 2 outlined in the Results section.

D. Receptor for photoinduction

Expose ten sixty-day-old *Xanthium* plants (two/treatment) to the light and dark requirements indicated in Table 44–3 for five cycles. Then return all plants to long-day conditions.

TABLE 44–3 Receptor for photoinduction

Pot Number	Condition of Plants	Light (hr/cycle)	Dark (hr/cycle)	Number of Cycles (days)	Class Results†	
					1	2
D-1	All leaves removed	8	16	5		
D-2	All leaves removed except one (No. 3 leaf)*	8	16	5		
D-3	Intact plant	8	16	5		
D-4	Intact plant	24	0	5		
D-5	Intact plant: Cover one leaf (No. 3) with a light-tight aluminum foil envelope from 5:00 P.M. to 9:00 A.M. each day.	24	0	5		

* Number 3 leaf should be approximately 7 to 9 cm in length from the base of the mid rib to the tip and is the third largest leaf after starting with the first smallest leaf which measures longer than 1 cm.
† Use an average value for class results as determined by Methods 1 or 2 outlined in the Results section.

*Results/
Conclusions*
Photoinduction of flowering in *Xanthium* may be determined by either one of the following methods or by subdividing the replicates for each treatment and using both methods. Remember, the class results should be pooled for adequate interpretation.

    1.   Measurement of the flowering stages of the apical bud: Ten days after the beginning of the experiment, examine the bud of each plant under a binocular dissecting microscope and record the stage of development according to the diagram and criteria given in Figure 44–1. Use the stages indicated as a guideline, and for any observed intermediate stage make your own

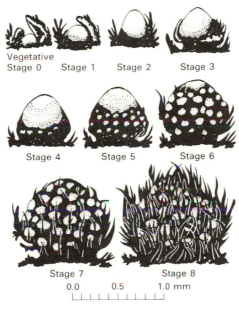

Vegetative
Stage 0   Stage 1   Stage 2   Stage 3

Stage 4   Stage 5   Stage 6

Stage 7   Stage 8

0.0   0.5   1.0 mm

Floral Stage	Criterion
0	Vegetative. Shoot apex relatively flat and small.
1	First clearly visible swelling of the shoot apex.
2	Floral apex at least as high as broad, but not yet constricted at the base.
3	Floral apex constricted at the base, but no flower primordia yet visible.
4	First visible flower primordia, covering up to the lower one-quarter of the floral apex.
5	Flower primordia covering from one to three-quarters of the flower apex.
6	Flower primordia covering all but the upper tip of the floral apex.
7	Floral apex completely covered by flower primordia. Slightly to moderately pubescent.
8	Very pubescent and showing some differentiation of flower parts. At least one millimeter basal diameter.

FIGURE 44–1   Flower stages and their criteria for the apical bud. (Reprinted with permission of the Macmillan Company from *The Flowering Process* by Frank B. Salisbury. Copyright © 1963, The Macmillan Company.)

interpolation and round off to the nearest whole number. Indicate the class results in the tables provided.

2.   Percentage of flowering plants for each treatment: Four to five weeks after photoinduction, examine the plants and record the percentage of flowers per plant and/or percentage of flowering plants for each treatment. Record the results of the entire class.

Present your conclusions concerning the critical day length, the number of inductive cycles necessary for flowering of *Xanthium*, and the receptor for photoinduction. Also explain whether *Xanthium* is classified as a short-day or long-day flowering plant. In addition to the general conclusions drawn from this exercise, present the current general ideas that relate to flower initiation in long-day and short-day flowering plants.

Acknowledgment   This exercise was adapted for class use from the work of Frank B. Salisbury.

References

Bonner, J. and J. A. D. Zeevaart. 1962. Ribonucleic acid synthesis in the bud, an essential component of floral induction in *Xanthium*. *Plant Physiol.* 37: 43–49.

Brady, J. (ed.). 1982. *Biological Timekeeping*. Society for Experimental Biology Seminar, Series 14. Cambridge: Cambridge University Press.

Chailakhyan, M. Kh. 1968. Internal factors of plant flowering. *Ann. Rev. Plant Physiol.* 19: 1–36.

Gardner, W. W. and H. A. Allard. 1920. Effect of the relative length of the day and night and other factors in growth and reproduction in plants. *Agric. Res.* 18: 553–606.

Hendricks, S. B. 1964. Photochemical aspects of plant photoperiodicity. In Giese, A. C. (ed.), *Photophysiology*. New York: Academic Press, pp. 305–331.

Hendricks, S. B. and H. A. Borthwick. 1954. Photoperiodism in plants. *Proc. Intern. Photobiol. Cong.* (1st), pp. 23–35.

Hillman, W. S. 1962. *The Physiology of Flowering*. New York: Holt, Rinehart and Winston, Inc.

Lincoln, R. G., D. L. Mayfield, and A. Cunningham. 1961. Preparation of a floral initiating extract from *Xanthium*. *Science* 133: 756.

Lockhart, J. A. 1961. Mechanism of the photoperiodic process in higher plants. In Ruhland W., (ed.), *Encyclopedia of Plant Physiology* 16: 390–438.

Salisbury, F. B. 1955. The dual role of auxin in flowering. *Plant Physiol.* 30: 327–334.

Vince-Prue, D. 1983. Photomorphogenesis and flowering. In Shropshire, W., Jr. and H. Mohr (eds.), *Encyclopedia of Plant Physiology*. (New Series), Vol. 16B. Berlin: Springer-Verlag, pp. 457–490.

EXERCISE 45

Photoreversible control of *Avena* coleoptile elongation

During the early research relating to the extraction and characterization of phytochrome, it was deemed important to develop assay systems that made it possible for scientists to observe phytochrome transformations *in vivo* without interference from protochlorophyll.

Hopkins and Hillman in 1965 (see the references at the end of this exercise) reported that the apical segments of *Avena* coleoptiles might be used as a system to study phytochrome-related growth. They found that the elongation of apical segments of *Avena* coleoptiles, incubated in distilled water or a phosphate-sucrose medium, was promoted by exposure to low-intensity red light. Further, the elongation response to red light treatment was primarily due to increased cellular size and was almost completely nullified if the red light treatment was followed immediately by exposure to far-red light. Thus, the elongation of *Avena* coleoptile segments represented a good assay to study the red, far-red reversible photomorphogenic pigment phytochrome. Although the system was not extensively used in phytochrome studies, it is provided here because it illustrates the effect of light on a phytochrome-mediated growth system that is relatively easy to perform and evaluate.

Purpose

To demonstrate the red, far-red, photoreversible behavior of a phytochrome-dependent growth response in *Avena* (oat) coleoptiles.

Materials

130 oat seedlings (*Avena sativa* 'Clintland'), three days old

40 ml 0.1M phosphate buffer stock (10.9 g KH_2PO_4 and 2.8 g Na_2HPO_4 dissolved in and diluted to 1 liter with distilled water)

1.5 g sucrose

12 beakers (50 ml)

12 3 in. pieces of aluminum foil

razor blade

rule (mm)

small spatula or forceps

red light source (see Appendix II)

far-red light source (see Appendix II)

dissecting microscope

Procedure

A. **Preparation of seedlings**

Soak approximately 50 to 75 ml of oat seeds in twice their volume of distilled water containing 0.05% Tween-20 (v/v) for thirty minutes. While the seeds are soaking, fill a polyethylene vegetable crisper (approximately 12 × 8 × 4 in.) with about 50 g vermiculite. Moisten the vermiculite with 600 ml water and thoroughly mix.

When the seeds are ready for planting, drain the seeds and scatter them uniformly over the surface of the vermiculite. Cover the basin to prevent evaporation and place it in a darkroom at 25°C. The seedlings will be ready for use seventy-two hours after sowing.

B. **Buffer-sucrose medium**

Dilute 40 ml of the stock buffer (0.1M phosphate buffer) to 200 ml with distilled water. This five-fold dilution should have a pH of approximately 6.0 to 6.4. Dissolve 1.5 g sucrose in 100 ml of the diluted buffer. Now dispense 5 ml of the buffered sucrose solution into each of twelve numbered beakers. Cover the beakers with aluminum foil.

C. **Coleoptile segment cutting, irradiation, and incubation**

All the following procedures involving the handling of plant materials should be performed in darkness or under a dim green safelight unless contra-indicated.

Working in the darkroom (where the seedlings were grown) under a dim green safelight, excise the apical 5 mm of the coleoptile from 130 seedlings. To perform this procedure, first remove a few seedlings from the vermiculite, pulling off excess vermiculite and separating the seedlings with your fingers. Then lay the seedlings on the mm rule so that the apex is aligned with a convenient mark. Using a razor blade, cut the coleoptile 5 mm below the apex. As each 5 mm segment is cut, place it in an open petri dish containing some of the diluted buffer without sucrose. Cut only one segment from each seedling.

When you have collected 130 segments, transfer ten segments to each of the twelve beakers. Then irradiate the segments according to the following schedule:

1. Expose beakers 3, 4, 7, 8, 9, 10, 11, and 12 to 5 min. of red light.
2. Expose beakers 5 through 12 to 5 min. far-red light.
3. Expose beakers 9 through 12 to 5 min. red light.
4. Expose beakers 11 and 12 to 5 min. far-red light.

During the individual light treatments, be sure that all those beakers not to be exposed at that time are safely maintained in total darkness (in a drawer or light-tight cabinet). After the last light treatment, cover all the beakers again and store them in complete darkness overnight (about eighteen to twenty hours is best).

Measure the lengths of the ten segments remaining in the petri dish with the aid of the dissecting microscope and estimate their lengths to the nearest 0.5 mm. This operation may be performed in the laboratory. Average the length for all ten segments and record this value as the initial length of the segments.

D. Segment measurements after irradiation and incubation

After the treated and control segments have incubated for the prescribed time, remove them to the laboratory and measure them with the aid of the dissecting microscope. Determine the growth increment for each one of the treatments according to the following expression:

$$\text{growth increment} = L_2 - L_1$$

where L_2 represents the length of an individual experimental segment after incubation and L_1 represents the average length of the segments determined from the ten segments measured at the beginning (L_1 will be the same value in all calculations).

Also calculate the total and average growth increment for each treatment and standard errors according to the procedure outlined in Table 45–1.

Measurements and calculations:

Results/ Conclusions

In addition to Table 45–1, illustrate the results in the form of a bar graph indicating the average growth increments on the ordinate against the various light treatments on the abscissa.

Interpret your results on the basis of the red and far-red photoreversible behavior of the phytochrome-dependent growth response studied.

TABLE 45–1   Standard error of the mean growth increments

Parameter	Symbol	Determination	Treatment Number											
			1	2	3	4	5	6	7	8	9	10	11	12
Number of segments per treatment	N	—	10	10	10	10	10	10	10	10	10	10	10	10
Sum of growth increments (mm)	$\Sigma\chi$	$\Sigma\chi = \chi + \chi + \cdots$												
Mean of growth increment/treatment	$\bar{\chi}$	$\bar{\chi} = \dfrac{\Sigma\chi}{N}$												
Sum of individual growth increments squared	$\Sigma(\chi^2)$	$\Sigma(\chi^2) = \chi^2 + \chi^2 + \chi^2 + \cdots$												
Sum of increments squared	$(\Sigma\chi)^2$	$(\Sigma\chi)^2 = (\Sigma\chi)(\Sigma\chi)$												
Correction factor	CF	$CF = \dfrac{(\Sigma\chi)^2}{N}$												
Sum of individual growth increments from the mean squared	Σx^2	$\Sigma x^2 = \Sigma(\chi^2) - CF$												
Standard error of the mean	$S\bar{x}$	$S\bar{x} \sqrt{\dfrac{\Sigma x^2}{N(N-1)}}$												

Acknowledgment   This exercise was adapted for class use by W. G. Hopkins, Bryn Mawr College.

References

Brady, J. (ed.). 1982. *Biological Timekeeping*. Society for Experimental Biology Seminar, Series 14. Cambridge: Cambridge University Press.

Hopkins, W. G. and W. S. Hillman. 1965. Phytochrome changes in tissues of dark-grown seedlings representing various photoperiodic classes. *Amer. J. Botany* 52: 427–432.

Hopkins, W. G. and W. S. Hillman. 1965. Response of excised *Avena* coleoptile segments to red and far-red light. *Planta* 65: 157–166.

Mohr, H. 1978. *Lectures on Photomorphogenesis*. Berlin: Springer-Verlag.

Shropshire, W. and H. Mohr. (eds.). 1982. Photomorphogenesis. In Shropshire, W., Jr. and H. Mohr (eds.), *Encyclopedia of Plant Physiology*, New Series, Vol. 16, Berlin: Springer-Verlag.

Smith, H. (ed.). 1976. *Light and Plant Development*. London: Butterworth.

Vince-Prue, D. 1975. *Photoperiodism in Plants*. London: McGraw-Hill and Co.

EXERCISE 46

Growth and development

Introduction

Many plants display volume changes in various cells and organs daily; yet many of these changes are temporary and not representative of true growth. To the biologist, growth involves the formation of new cells and/or increase in cellular size due to water uptake, followed by synthesis of cell walls and other components, thus making the changes irreversible. Whether a given cell grows or not depends largely upon whether the cell wall is elastic enough to allow for increase in cell volume. If so, water coming into the cell will stretch it, and permanent changes in volume will follow. It is equally important to recognize that the growth of plants proceeds by the addition of new cells to the plant body in specific regions of growth (the meristems). Ultimately, growth of plants is a combination of cell division and cell expansion.

Early in the history of modern plant physiology, Charles Darwin discovered evidence that plants were capable of responding to differential lighting by bending toward the source of light. Many years later, in this century, Dr. Fritz Went, continuing Darwin's studies of light responses in oats, isolated a substance he called auxin. Since that time, auxin has been shown to be a very significant plant hormone that functions in numerous aspects of plant growth and development.

Auxins are found in particularly large amounts at the growing regions of plants. They have been shown to have effects on root initiation, root growth, stem growth, and many other plant processes. Additionally, it has been found that in higher levels than that normally used on plants certain synthetic analogs of auxin can kill plants. This action, however, has been found to be selective toward certain types of plants. The result has been the development of chemicals that have had widespread application in agriculture for weed control.

In this exercise, you will set up several experiments that will provide information on how plants grow as well as how auxin affects their growth.

Purpose

To demonstrate growth in pea roots and stems, the effects of auxin on root initiation, and the herbicidal action of auxins on plants.

Materials

pea (*Pisum sativum* L.) seedlings, three to four days old and seven to ten days old (see Procedure, Parts A, B, and C)

bean (*Phaseolus vulgaris* L.) seedlings, seven to ten days old (see Procedure, Parts A and D)

rye and cucumber seeds

petri dishes (9 cm)

filter paper (9 cm)

paper towels

ruler

marker

beakers or flasks (250 ml, 500 ml)

solutions of indoleacetic acid in concentrations of 0.01, 0.10, and 1 mg/l, respectively

petri dishes containing two pieces of filter paper and sterilized

solutions of 2,4-dichlorophenoxyacetic acid in concentrations of 0.00001, 0.01, 1, and 100 mg/l, respectively

Procedure

A. Plant material

To obtain seedlings of the desired stage (three- to four-day-old seedlings), sterilize pea seeds by immersing them for ten to fifteen minutes in commercial hydrochlorite solution (Clorox) that has been diluted 1:10 with distilled water. Then rinse the seeds in distilled water.

After sterilization, place the seeds on moist vermiculite in flats and cover with approximately one-half inch of additional vermiculite. The seeds should readily germinate at room temperature or in the greenhouse. To maintain the germinating seedlings, add distilled water daily or when needed to keep the vermiculite moistened.

Flats of pea seedlings (seven to ten days old) may be prepared and grown in the same manner as described above. For Part D, soak garden bean seeds (*Phaseolus vulgaris* L.) in Clorox (1:10) for ten minutes. Rinse the seeds and place them in a 1000 ml Erlenmeyer flask containing 500 ml of distilled water with aeration for twenty-four hours at room temperature. Rinse the seeds and sow in a tray or flat containing vermiculite. Cover the seeds with one-half inch of vermiculite, water, and incubate in a growth chamber or greenhouse for seven to ten days.

B. Growth of pea roots

Carefully remove five three- to four-day-old pea seedlings from the flat in the laboratory. Select seedlings that have primary roots 1 or 2 cm long. Place several layers of paper towels or filter paper in the bottom of a 9 cm petri dish. The layer should be thick enough so that when the seedlings are spread on it, addition of the petri dish lid will press against the cotyledons and hold the seeds firmly in place. Soak the paper layer with water, pouring off any excess. Place five seedlings in the dish with all the roots pointing in the same direction. Make marks across the roots at 2 mm intervals, from the tip, using a ruler and the marking pen provided by the instructor. Measure and record the length of each root to the nearest millimeter. Cover the dish and secure the lid with a rubber band or tape. Store the dish in an upright position with the roots pointing downward. Two to three days later, the roots may be measured again or the dishes may be stored in the refrigerator temporarily before your final measurements. Be sure to record the initial and final lengths of the pea roots and the average length of the intervals in millimeters.

C.   Growth of pea stems

Select two pea seedlings (seven to ten days old) from the flats provided. The number of plants used will depend upon the instructor and the facilities available. Carefully remove each plant from the flat so as to reduce the amount of damage to the roots. Plant both in a single pot of fresh vermiculite. With a marker, place thin marks at 3 mm intervals from the soil level, along the stem to the shoot apex of one of the plants. Record the height of each plant from the soil level to the shoot apex. Place the potted plants in the greenhouse and take final measurements approximately one week later. Be sure to record the initial and final lengths of the pea shoots in millimeters.

D.   Effects of auxins on root initiation

Cut off the roots of three to five bean plants and place the stems into each beaker or flask containing the following solutions of indoleacetic acid (IAA):

Solution 1: distilled water only (control)

Solution 2: IAA (0.01 mg/l)

Solution 3: IAA (0.10 mg/l)

Solution 4: IAA (1 mg/l)

Allow the plants to stand in the greenhouse or growth chamber for a week. At that time examine each of the beakers and score the plants for relative number of roots (0 = no roots, + = 1–5, ++ = 6–50, +++ = 50–150, ++++ = over 150). Be sure to record the rooting score (average of all plants in the flask or beaker) against the appropriate auxin concentration and control.

E.   Herbicidal action of auxins

You will be provided with petri dishes, each containing two pieces of filter paper. Do not open these repeatedly, as they have been sterilized. Keep each dish open as short a time as possible and to each quickly add 10 ml of one of the 2,4-dichlorophenoxyacetic acid solutions provided. The concentrations of 2,4-D are:

Solution 1: distilled water (control)

Solution 2: 2,4-D (0.00001 mg/l)

Solution 3: 2,4-D (0.01 mg/l)

Solution 4: 2,4-D (1 mg/l)

Solution 5: 2,4-D (100 mg/l)

You should have one dish for each solution. Place fifteen rye seeds in half of the dish and fifteen cucumber seeds in the other half of the dish. Store dishes for five to seven days at room temperature and in normal room light (no special lighting conditions are required). During the next lab period, measure the length of the primary root of each seed and record the average length for each treatment. *Note:* Be sure to wash your hands thoroughly after handling all treated seeds.

Results/ Conclusions

Interpret the results for all parts of the exercise. In addition to your concluding remarks, answer the following questions:

1. Are there regional differences in the growth of roots and shoots?

2. Does growth in stems generally occur over larger distances than growth in roots?

3. Which solution of IAA seemed to produce the largest roots? What do the results tell you about the levels of IAA that should be used to promote rooting of cuttings?

4. Is 2,4-D always an inhibitor of root growth?

5. In what way does 2,4-D affect the growth of rye and cucumber differently?

6. What is the importance of the control (distilled water only) in Parts D and E?

Acknowledgment This exercise was adapted for class use by Michael Strauss, Northeastern University.

References

Cleland, R. 1971. Cell wall extension. *Ann. Rev. Plant Physiol.* 22: 197–222.

Darwin, C. 1881. *The Power of Movement in Plants.* New York: D. Appleton.

Firn, R. D. and J. Digby. 1980. The establishment of tropic curvatures in plants. *Ann. Rev. Plant Physiol.* 31: 131–148.

Leopold, A. C. and P. E. Kriedemann. 1975. *Plant Growth and Development,* 2nd edition. New York: McGraw-Hill.

Moore, T. C. 1979. *Biochemistry and Physiology of Plant Hormones.* New York: Springer-Verlag.

Scott, T. K. 1972. Auxins and roots. *Ann. Rev. Plant Physiol.* 23: 235–258.

Thimann, K. V. 1977. *Hormone Action in the Whole Life of Plants.* Amherst: University of Massachusetts Press.

Torrey, J. G. and D. T. Clarkson (eds.). 1975. *The Development and Function of Roots.* New York: Academic Press, Inc.

Wareing, P. F. and I. D. J. Phillips. 1978. *The Control of Growth and Differentiation in Plants,* 2nd edition. New York: Pergamon Press.

Went, F. W. 1974. Reflections and speculations. *Ann. Rev. Plant Physiol.* 25: 1–26.

EXERCISE 47

Phototropism

Introduction

When some plant parts such as coleoptiles, leaves, stems, or flowers are illuminated by unilateral light, they often exhibit a growth response involving a bending toward the light (*phototropism*). The bending is caused by cells on the shaded side becoming elongated at a greater rate than cells on the illuminated side. This differential growth response seems to be due to an unequal distribution of auxin favoring the shaded side. According to the Cholodny-Went theory, it is the higher concentration of auxin on the shaded side that results in the differential growth response indicated by greater elongation on the shaded side.

The unequal distribution of auxin, as induced by unilateral light, has intrigued plant scientists for years. Several explanations for the unequal distribution of auxin are the light-induced inactivation of auxin on the light side, light-induced lateral transport of auxin, or inhibition of basipetal transport of auxin to or on the light-illuminated side of the plant part in question. The current evidence favors the explanations that either light-induced lateral transport of auxin or inhibition of basipetal transport are likely mechanisms for auxin distribution in stems and coleoptiles.

Purpose

To illustrate the influence of the apex and indoleacetic acid upon the phototropic response of *Avena* coleoptiles.

Materials

70 oat seeds (*Avena sativa* L. 'Victory'): Most of the commonly available varieties will serve, but greater sensitivity is usually obtained with late-maturing cultivars such as Victory, 20th Century, Anthony, and White Russian.

100 ml Tween-20 aqueous solution (0.05%, v/v)

2 battery jars, containing a layer of moist cotton on the bottom

aluminum foil covers (5 mm long)

2 cardboard cartons (large enough to close when a battery jar is placed inside)

cutting device (two double-edge razor blades with one edge imbedded into a cork stopper so the blades are separated by a measured interval)

indoleacetic acid in lanolin (1%, w/w)

light source (40 watt bulb)

Procedure   **A.   Seed germination**

Soak seventy oat seeds in twice the volume of distilled water containing 0.05% Tween-20 (v/v) for sixty minutes. Transfer thirty of the soaked seeds to a layer of moist cotton covering the bottom of a battery jar (No. 1). Transfer the remaining forty seeds to another battery jar (No. 2) also containing a layer of wetted cotton. Cover the jars and place in a darkroom or cabinet and maintain in the dark until the coleoptiles have reached a height of approximately 2 cm.

B.   Phototropism and the coleoptile tip

In a darkroom under dim red light, treat the coleoptiles of the seedings in battery jar No. 1 according to the following steps (it is possible to accomplish the operation under very subdued light in the laboratory if necessary):

1.   Cover the tips of ten coleoptiles with small covers of aluminum foil about 5 mm in length.

2.   Amputate the apical 2 to 3 mm of ten coleoptiles. Leave the remaining ten seedlings intact.

3.   Cover the jar and place it near the center of one end of a cardboard carton of suitable size, the inner walls of which have been painted a flat black. Seal the carton with gummed paper tape so that it is light-proof.

4.   Cut a small hole (about 1 cm in area) about 10 cm from the bottom of the carton in the end opposite to that close to the battery jar. Orient the carton in such a manner that light will enter the hole.

5.   Forty-eight hours after the continual unilateral light treatment, open the carton and examine the coleoptiles. Observe their position relative to the incident light.

Observations:

C.   **Effect of indoleactic acid (IAA) upon phototropism:**

Perform the following steps in a darkroom under red light (if a darkroom is not available, the following steps may be performed under subdued light conditions for a short period of time):

1.   Amputate the apical 2 to 3 mm of thirty coleoptiles. Using a mixture of 1% IAA in lanolin (w/w), make a streak of the mixture from the tip to the base and on the same side of ten decapitated coleoptiles (the side that will be away from the unilateral light). To ten other decapitated coleoptiles, smear the IAA on the side toward the light source. Leave ten decapitated coleoptiles untreated and the remaining ten seedlings intact.

2.   Transfer the battery jar to a carton and expose the seedlings to unilateral light as in Part B.

3.   Forty-eight hours after the light treatment, open the carton and examine the coleoptiles. Note their position relative to the incident light.

Observations:

Results/ Conclusions

Illustrate the results with a suitable table and give your interpretations of the auxin theory of phototropism. Is this theory an adequate explanation of all phototropic reactions?

References

Dennison, D. S. 1984. Phototropism. In Wilkins, M. B. (ed.), *Advanced Plant Physiology.* London: Pitman Publishing, Limited, pp. 149–162.

Firn, R. D. and J. Digby. 1980. The establishment of tropic curvature in plants. *Ann. Rev. Plant Physiol.* 31: 131–148.

Darwin, C. 1881. *The Power of Movement in Plants.* New York: The Appleton Co.

Galston, A. W., P. J. Davies, and R. Sater. 1980. *The Life of the Green Plant,* 3rd edition. Englewood Cliffs, N.J.: Prentice-Hall, Inc.

Gressel, J. and B. Horwitz. 1982. Gravitropism and phototropism. In Smith, H. and D. Grierson (eds.), *Molecular Biology of Plant Development. Berkeley: University of California Press,* pp. 405–433.

Jacobs, W. 1979. *Plant Hormones and Plant Development.* New York: Cambridge University Press.

Laetsch, W. M. and R. E. Cleland (eds.). 1967. *Papers on Plant Growth and Development.* Boston: Little, Brown and Co.

Leopold, A. C. and P. Kriedemann. 1975. *Plant Growth and Development,* 2nd edition. New York: McGraw-Hill Book Co.

Moore, T. 1979. *Biochemistry and Physiology of Plant Hormones.* New York: Springer-Verlag.

EXERCISE 48

Gravitropism

If a plant or intact seedling is placed in a horizontal position so that the stem and root axes are parallel to the earth's gravitational field, the stem will grow upward away from the gravitational force while the root system will grow downward following gravity. Accordingly, the stem exhibits *negative gravitropism* and the root exhibits *positive gravitropism*. (The term *geotropism* is also commonly used.) These responses are similar for developing seedlings and plants under normal conditions with the normal growth of the plant resulting in stem and root growth perpendicular to the gravitational field.

Cholodny and Went originally proposed the idea that the differential growth response exhibited by a horizontally placed stem or root is due to the accumulation of auxin (IAA) in the lower side in response to the force of gravity. In horizontally placed stems, the accumulation of auxin on the lower side causes accelerated growth on the lower side, resulting in stem curvature upward (negative gravitropism). Further, the Cholodny-Went theory states that roots are more sensitive to auxin than stems. Therefore, the higher concentration of auxin toward the lower sides of roots inhibits growth (elongation) of the lower cells. The concentration of the auxin on the upper side may be reduced to the stimulatory level, resulting in accelerated growth of the upper cells. The differential elongation response (some elongation of the top cells) causes bending of the organ to gravity.

The Cholodny-Went theory, as it applies to the negative gravitropism of stems, may be essentially correct. However, there are currently different views concerning the role of auxin and growth inhibitors in the gravitropism of roots.

There seems to be a good deal of evidence that auxin indole-3-acetic acid (IAA) is transported in an acropetal direction in roots, and that growth inhibitor(s) produced in the root cap is transported basipetally into the elongation zone of the root. Normally, the inhibitor (possibly abscisic acid) is distributed throughout the cells of the root in the zone of elongation. IAA moving acropetally into the zone stimulates elongation over any growth inhibition by the dilute inhibitor. However, under the influence of gravity (when the root is placed in a horizontal position), the inhibitor accumulates in the lower cells and in high concentrations inhibits elongation. Thus growth of the upper cells is accelerated by the auxin present, while the growth of the lower cells is inhibited, resulting in a bending response to gravity.

Purpose

To study gravitropism in corn seedlings and the role of the root tip.

Materials

50–60 germinating corn grains (see Procedure, Part A)	2 battery jars
	soil (sufficient amount to fill each battery jar)
wide-mouthed bottle	
cheesecloth	aluminum foil

Procedure

A.   Seed germination

Place approximately fifty to sixty corn grains in a wide-mouthed bottle and soak them in tap water for several hours. Drain off the water and rinse several times with tap water. Then cover the opening of the bottle with several layers of cheesecloth and secure them to the bottle with string. Invert the bottle at an angle in a sink out of direct sunlight. Rinse the seeds with tap water several times each day until some of the coleoptiles are 1 to 2 cm long. Keep the rest for a longer period for the seedlings to develop roots about 3 cm long.

B.   Gravitropic response

Fill a battery jar half full of moist soil and introduce twenty of the germinating grains (coleoptiles 1 to 2 cm long) so that the root tip and coleoptile are visible through the glass. Arrange the grains in different positions so that one-third have the root ends pointed downward, another third the root end pointed upward, and the remaining seeds with their root ends in intermediate positions.

Fill the remainder of the battery jar with moist soil and stand it in diffuse light. Cover the sides of the jar with aluminum foil and keep covered between observations. Observe the behavior of the roots and the coleoptiles as the plants grow.

Observations:

C.   Root tip and gravitropic curvature

Select twenty corn seedlings with roots approximately 3 cm in length. Cut off the terminal 2 mm of root tip from each of ten seedlings (use a sharp razor blade).

Fill half a battery jar with moist sand and insert the seedlings so the roots are between the sand and the glass. Arrange the grains so that the roots are oriented in a horizontal position. Arrange ten intact germinating seedlings (roots 3 cm long) in the same manner and fill the battery jar with moist sand. Cover the battery jar with aluminum foil between observations. Observe the behavior of the root tips from time to time over a period of several days.

Observations:

Results/ Conclusions

Account for any growth differences observed between the intact seedling roots and the seedlings with decapitated root tips. What is the major theory to explain the role of the root tip and auxins in gravitropism? Can you suggest newer theories?

Acknowledgment

This exercise was adapted from B. S. Meyer, D. B. Anderson, and C. A. Swanson. 1955. *Experimental Plant Physiology,* 3rd edition. D. Van Nostrand Co. Inc., Princeton, N.J., pp. 160–162.

References

Audus, L. J. 1975. Geotropism in roots. In Torrey, J. and D. T. Clarkson (eds.), *The Development and Function of Roots.* London: Academic Press, pp. 327–363.

Darwin, C. 1881. *The Power of Movement in Plants.* New York: The Appleton Co.

Digby, J. and R. D. Firn. 1979. An analysis of the changes in growth rate occurring during the initial stages of geocurvature in shoots. *Plant, Cell and Environment* 2: 145–448.

Firn, R. D. and J. Digby. 1980. The establishment of tropic curvatures in plants. *Ann. Rev. Plant Physiol.* 31: 131–148.

Jacobs, W. P. 1979. *Plant Hormones and Plant Development.* Cambridge: Cambridge University Press.

Moore, T. C. 1979. *Biochemistry and Physiology of Plant Hormones.* New York: Springer-Verlag.

Went, F. W. and I. V. Thimann. 1937. *Phytohormones*. New York: Macmillan Co.

Wilkins, M. B. 1984. Gravitropism. In Wilkins, M. B. (ed.), *Advanced Plant Physiology*. London: Pitman Publishing, Limited, pp. 163–185.

Avena coleoptile section test: the basis for a bioassay

Introduction

The *Avena* coleoptile section test is based on auxin stimulation of cellular elongation. Unlike the *Avena* coleoptile curvature test, the section test does not involve transport of auxin or differential growth. The *Avena* section test measures the effect of auxin over a wide range of concentrations and is not hindered by problems of transport of growth regulators as is the curvature test.

The growth response of coleoptile sections is directly proportional to the logarithm of the concentration of auxin used. In contrast, the *Avena* coleoptile curvature test is directly proportional to the amount of auxin tested. The curvature test, therefore, is much more sensitive, but unlike the section test, is confined to a narrow concentration range. The section test was used widely in the past due to its simplicity of performance and ease of interpretation. The procedures are outlined in the following exercise.

Purpose

To illustrate the techniques used in one bioassay for auxins and to study the effect of indole-3-acetic acid on the growth of *Avena* coleoptile sections.

Materials

50 oat seedlings (*Avena sativa* L. 'Victory'), grown in the dark and ready before the scheduled class meeting (see Procedure, Part A)

100 ml sodium hypochlorite solution (1%)

100 ml KH_2PO_4 (0.01M) buffer solution (pH 4.5) containing 2% (w/v) sucrose

darkroom (25 to 26°C and about 85% relative humidity)

lamp (equipped with 10 watt red bulb)

10 petri dishes

cutting tool (a double-bladed cutting tool may be constructed with two blades held parallel to each other and separated by a 6 mm wedge)

forceps

10 microscope slides

dissecting microscope

ruler (graduated to 0.1 mm)

Procedure   **A.   Seed germination and seedling growth**

Sterilize sixty oat seeds (*Avena sativa* L. 'Victory') for five minutes in 1% sodium hypochlorite solution. After thoroughly rinsing in water, soak the seeds in sterile distilled water for approximately one hour.

Sow the sterilized seeds in flats of vermiculite and grow in the dark at 25°C and a relative humidity of about 85%. The seedlings may also be germinated under continuous low-intensity red light to increase their sensitivity to auxin (use a lamp equipped with a 10 watt red bulb).

B.   Coleoptile section test

Note: The cutting operation should be performed in a darkroom under dim red light.

Seventy-two hours after sowing the seed (coleoptiles approximately 25 to 30 mm in length), prepare fifty uniform subapical sections (6 mm in length) of the coleoptiles with the double-bladed cutting tool. The upper cut of each section should be made exactly 3 mm below the tip of the selected coleoptiles.

Immediately after cutting, place the sections in a beaker containing a 2% (w/v) sucrose and 0.01M KH_2PO_4 buffer solution (pH 4.5). This solution will henceforth be referred to as *basal medium*. Soak the sections with basal medium for a minimum of one hour to wash out the endogenous auxin. After soaking, place five sections into each petri dish containing 20 ml of the following test solutions set up in duplicate:

Solution 1: basal medium

Solution 2: basal medium + IAA (10^{-7}M)

Solution 3: basal medium + IAA (10^{-6}M)

Solution 4: basal medium + IAA (10^{-5}M)

Solution 5: basal medium + IAA (10^{-4}M)

Incubate the segments in the solutions in darkness for ten hours, after which time the sections in one complete test series may be measured. Determine the length of the sections in the other set of solutions after a total incubation time of twenty hours.

After the prescribed incubation time, transfer the sections of each treatment to a microscope slide and add a drop of the incubation solution so that they will not dry out. With a dissecting microscope, equipped with an ocular micrometer or by using a graduated centimeter rule, determine the length of the sections to the nearest 0.1 mm.

Measurements:

Results/
Conclusions

Determine the average length of the segments for each treatment, including the average length of the sections incubated in basal medium alone (controls). To determine the amount of auxin-induced growth, subtract the mean length of the control segments from that of the auxin-treated segments. Plot the resulting differences against the logarithm of auxin concentration in the form of a dose response curve.

Concluding statements should be concerned with the effect of IAA on coleoptile section growth and any pertinent information relating to the specificity and sensitivity of the test.

References

Bandurski, R. S. and H. M. Nonhebel. 1984. Auxins. In Wilkins, M. B. (ed.), *Advanced Plant Physiology*. London: Pitman Publishing, Limited, pp. 1–20.

Bentley, J. A. 1950. An examination of a method of auxin assay using the growth of isolated sections of *Avena* coleoptiles in test solutions. *J. Exp. Bot.* 1: 201–13.

Bentley, J. A. and S. Housely. 1954. Bio-assay of plant growth hormone. *Physiol. Plantarum* 7: 405–420.

Cohen, J. D. and R. S. Bandurski. 1982. Chemistry and physiology of bound auxins. *Ann. Rev. Plant Physiol.* 33: 403–430.

McRae, D. H. and J. Bonner. 1953. Chemical structure and antiauxin activity. *Physiol. Plantarum* 6: 485–510.

Mitchell, J. W. and G. A. Livingston. 1968. Methods of studying plant hormones and growth-regulating substances. *Agric. Handbook No. 336*, USDA, pp. 23–25.

Nitsch, J. P. and C. Nitsch. 1956. Studies on the growth of coleoptile and first internode sections. A new, sensitive, straight-growth test for auxin. *Plant Physiol.* 31: 94–111.

Schneider, C. L. 1938. The interdependence of auxin and sugar for growth. *Amer. J. Botany* 25: 258–270.

Sirois, J. C. 1966. Studies on growth regulators. I. Improved *Avena* coleoptile elongation test for auxin. *Plant Physiol.* 41: 1308–1312.

Thimann, K. V. 1977. *Hormone Action in the Whole Life of Plants*. Amherst: University of Massachusetts Press.

EXERCISE 50

The split pea stem curvature test: the basis for a bioassay

Introduction

In the past, the split pea stem curvature test served as a bioassay to detect auxin in plant extracts.

In preparation for the test, stem sections of pea seedlings (variety 'Alaska') are slit longitudinally about two-thirds of their length. These split sections are floated on solutions containing known amounts of IAA or unknown solutions. When auxin is present, the epidermal cells respond to it with considerable growth in length. The cortical cells, however, exposed by the splitting of the sections, do not grow in length appreciably (in fact, they are inhibited), resulting in a differential growth response of the section. Consequently, after a suitable incubation period in a physiological concentration of auxin, positive curvature may be observed in which the two split ends curve toward the center of the slit and the split pieces bow outward. Within a certain concentration range, the response of the slit halves of the stem is roughly proportional to the logarithm of the concentration of auxin used.

As with other bioassays used in the past, this test, although not widely used to detect auxins in plant extracts, provided some of the early information about the physiological activity of auxin in plants.

Purpose

To study another method of bioassay for auxin and to illustrate the differential growth response of split pea stems.

225

Materials

30 pea seedlings (*Pisum sativum* L. 'Alaska') grown in the dark for eight to ten days (see Procedure, Part A)

4 separate indole-3-acetic acid solutions (30 ml of each at IAA concentrations of 10^{-2}M, 10^{-4}M, 10^{-5}M, and 10^{-6}M)

darkroom (temperature approximately 25 to 28°C)

lamp (equipped with 10 watt red bulb)

1 beaker (80 ml)

5 petri dishes

razor blades (a double-bladed cutting tool may be constructed with two blades held parallel to each other and separated by a wedge of desired thickness)

Procedure

A. Preparation of plant material

Note: Complete preparation before the scheduled class meeting.

Soak pea seeds (*Pisum sativum* L. 'Alaska') in running tap water for six hours. Then sow the seeds in flats of sand and place in total darkness for approximately eight to ten days. Be sure to water the seedlings periodically.

B. Section cutting

Harvest the seedlings in the laboratory and cut uniform sections (about thirty in number and 3 to 4 cm in length) from just below the top of the third internode.

Using a razor blade, slit each segment (about three-fourths down the length) and drop it in distilled water. Allow the split sections to soak in water for approximately thirty minutes to remove some of the endogenous auxin.

C. Response of split pea stem sections to indole-3-acetic acid

Place five split stem sections into petri dishes containing 20 ml of the following (each solution should be adjusted to pH 5.0):

Solution 1: distilled water
Solution 2: IAA (10^{-2}M)
Solution 3: IAA (10^{-4}M)
Solution 4: IAA (10^{-5}M)
Solution 5: IAA (10^{-6}M)

If time permits and at the discretion of the instructor, several synthetic auxins or other growth substances (cytokinins, gibberellins) may be tested. Auxin concentrations between 10^{-2}M and 10^{-6}M should be used.

Incubate the sections in all of the test solutions in the dark for a minimum of six hours. Then observe the results. If necessary the sections may be incubated for a longer period since the initial response should not change appreciably after six hours.

Observations:

Results/ Conclusions

Present your observations on a representative diagram of the segments incubated in distilled water and the various auxin solutions. It is possible to estimate the results quantitatively and, if facilities are available for making shadow graphs of the sections, follow the techniques outlined in some of the references given. In concluding statements, discuss the nature of the response and the theoretical basis of the test.

Acknowledgment

This exercise was adapted for class use from the work of F. W. Went.

References

Mitchell, J. W. and G. A. Livingston. 1968. Methods of studying plant hormones and growth-regulating substances. *Agric. Handbook No. 336*, USDA, pp. 32–34.

Thimann, K. V. and C. L. Schneider. 1938. Differential growth in plant tissues. *Amer. J. Bot.* 25: 627–641.

Went, F. W. 1937. On the pea test method for auxin, the plant growth hormone. *K. Akad. Wetenschap. Amsterdam Proc. Sect. Sci.* 37: 547.

Went, F. W. and K. V. Thimann. 137. *Phytohormones.* New York: Macmillan Co.

(Also see the references following Exercise 48.)

EXERCISE 51

Indoleacetic acid and adventitious root initiation

Introduction

Application of auxin to the severed end of a young stem (minus roots) stimulates the rate of root formation. As a result of this early observation, the commercial application of indole-3-acetic acid (IAA) and naphthalene acetic acid (NAA) to promote root formation in stem cuttings of economically useful plants has been a standard practice for years.

In the following exercise, Mung bean cuttings without roots will be exposed to solutions of IAA at varying concentrations. At the end of a five-day period, you will be asked to evaluate the results.

Purpose

To study the effect of indoleacetic acid on adventitious root initiation.

Materials

180 Mung bean cuttings (*Vigna radiata* L., R. Wilcz 'Berken'; see Procedure, Part A)

5 separate indole-3-acetic acid solutions (10 ml of each at IAA concentrations of:

- IAA 10^{-3}M ($10^{-5} \times$ gram molecular weight of IAA dissolved in 500 μl 95% ethanol and diluted with rapid mixing to 10 ml with distilled water): all distilled water used for reagents and treatments should be autoclaved.

- IAA 5×10^{-4}M (5 ml 10^{-3}M IAA diluted to 10 ml with distilled water)

- IAA 10^{-4} (1 ml 10^{-3} IAA diluted to 10 ml with distilled water)

- IAA 10^{-5} (1 ml 10^{-4} IAA diluted to 10 ml with distilled water)

- IAA 10^{-6} (1 ml 10^{-5} IAA diluted to 10 ml with distilled water)

10 ml distilled water containing 50 μl 95% ethanol

flat containing perlite

trays (29 × 18 × 5 cm)

growth chamber maintained at 77°F day and 73°F night, and sixteen hour photoperiod with about 55 μmol s^{-1}m^{-1} light

intensity supplied by a combination of fluorescent and incandescent lamps
18 shell vials (19 × 65 mm) and
shell vial holder

razor blades
commercial bleach
autoclaved distilled water

Procedure

A. Plant material

The cultivar 'Berken' has proven to be reliable; however, the cultivar 'Oriental Giant' can also be used successfully. The following technique is presented primarily to illustrate the effect of indoleacetic acid on adventitious root initiation. Auxins such as indolebutyric acid, naphthaleneacetic acid, or 2,4-dichloroacetic acid, can be substituted or compared with IAA.

Soak 100 ml of Mung bean seeds (*Vigna radiata* 'Berken') in 200 ml bleach solution (1:10 dilution) for ten minutes. Rinse the seeds and place in a 1000 ml Erlenmeyer flask containing 500 ml water and then aerate for twenty-four hours at room temperature (25°C). Rinse the seeds and sow in a tray containing 4 in. of perlite. Cover the seeds with a 1 in. layer of perlite, water with tap water, and germinate in a growth chamber for eight to nine days (see the list of materials). Five days after planting, fertilize with Hoagland's solution.

B. Indoleacetic acid treatment and the adventitious root initiation response

Uniform cuttings are made from eight- to nine-day-old seedlings. Cuttings are ready for assay when the primary leaves are fully expanded and the trifoliate bud has not expanded. Harvest the cuttings by severing the seedlings from their root systems and remove any attached cotyledons. Prepare the cutting so that each cutting consists of a 3 cm hypocotyl, epicotyl, primary leaves, and apical meristem and place in autoclaved distilled water. Select cuttings of uniform size to reduce variation.

Label the shell vials (three replicates per treatment) according to the following treatments:

Treatment 1: distilled water (control)

Treatment 2: IAA (10^{-3}M)

Treatment 3: IAA (5×10^{-4}M)

Treatment 4: IAA (10^{-4}M)

Treatment 5: IAA (10^{-5}M)

Treatment 6: IAA (10^{-6}M)

Dispense 1 ml of distilled water or the appropriate IAA solution in each appropriately labeled vial. Be sure to use a distilled water treatment that contains the same amount of ethanol used in the IAA solutions. Select ten cuttings at random and place in each shell vial. After the initial test solution is absorbed, about three hours, add sterile water and maintain the water level at the cotyledonary node during the five-day rooting period. After five days, count the roots and record the average number of roots per cutting.

Observations and measurements:

Results/Conclusions

Present a figure for the exercise in which the averages of the measurements made on the control and treated plants are illustrated. Is IAA translocated from the site of application to other areas on the treated cuttings? Is adventitious root initiation a polar process? In addition to stimulating adventitious root initiation, what other plant responses did the IAA influence? In additional remarks, discuss the site of IAA synthesis in plants and relate synthesis to site of action. What is the structure of IAA?

Acknowledgment

This exercise was adapted for class use by Dr. Charles W. Heuser, Department of Horticulture, The Pennsylvania State University.

References

Geneve, R. L. and C. W. Heuser. 1983. The effect of IAA, IBA, NAA, and 2,4-D on root promotion and ethylene evolution in *Vigna radiata* cuttings. *J. Amer. Soc. Hort. Sci.* 107: 202–205.

Geneve, R. L. and C. W. Heuser. 1983. The relationship between ethephon and auxin on adventitious root initiation in cuttings of *Vigna radiata* (L.) R. Wilcz. *J. Amer. Soc. Hort. Sci.* 108: 330–333.

Goeschel, J. D. and H. K. Pratt. 1968. Regulatory roles of ethylene in the etiolated growth habit of *Pisum sativum*. In Wightman, F. and G. Sutterfield (eds.), *Biochemistry and Physiology of Plant Growth Substances*. Ottawa: Range Press, pp. 1229–1242.

Haisig, B. E. 1971. Influence of indole-3-acetic acid on incorporation of ^{14}C-uridine by adventitious root primordia or brittle willow. *Bot. Gaz.* 132: 263–267.

Tripepi, R. R., C. W. Heuser, and J. C. Shannon. 1983. Incorporation of tritiated thymidine and uridine into adventitious-root initial cells of *Vigna radiata*. *J. Amer. Soc. Hort. Sci.* 108: 469–474.

Zimmerman, P. W. and F. Wilcoxon. 1935. Several chemical growth substances which cause initiation of roots and other responses in plants. *Contrib. Boyce Thompson Inst.* 7: 209–229.

EXERCISE 52

Rapid elongation of *Avena* coleoptiles

Introduction

Many of the observable biochemical physiological events due to auxin action occur soon after a plant part is exposed to it. Such responses are referred to as *rapid responses*.

Most of the studies of auxins on cell elongation are performed on excised plant material, such as *Avena* coleoptile sections or excised root sections that have little or no endogenous auxin supply. These plant materials are ideal for such study because the effect of exogenously applied auxin may be studied without interference from endogenous auxin. Coleoptiles are very sensitive to auxin and exhibit a response significantly greater than that in the absence of auxin.

Rapid auxin responses continue to be studied because they represent an important mechanism that accounts for adjustment of plants to environmental fluxes through a relatively rapid response. Also, scientists are able to obtain significant information about auxin action when the response is not far removed from the chemical reception of the hormone (that is, the time between reception and the response is relatively short).

Purpose

To measure rapid increase in growth by stacking coleoptiles held on thread, which amplifies the growth response so that increases in length can be measured in one hour and fifteen minutes or less.

Materials

phosphate buffer (pH 6.8), with 1 millimolar indoleacetic acid

Avena sativa 'Victory II' (enough seed so that each group can have at least twenty coleoptiles)

plastic basin, blackened, with lids

vermiculite

cotton thread

needle

2 square (1 cm) pieces of sheet plastic with hole in center

two-bladed cutter

block of paraffin

2 petri dishes (10 cm)

2 graduated pipets (10 ml)

binocular dissection microscope

cheesecloth and rubber band

beaker (400 ml)

red safelight (Wratten No. 1A)

darkroom

Procedure

Place *Avena sativa* 'Victory II' seed in a suitable beaker and cover with cheesecloth. Allow the seed to soak for two hours in running tap water. At that time, sow the seed in wetted vermiculite in plastic basins that have been painted black so that no light passes through. Then place the basins in a growth chamber at 26°C. Forty-eight hours after sowing, expose the coleoptiles for about one hour to dim red light (Wratten No. 1A). Five days later, the coleoptiles should be about one inch long (view in dim red light, Wratten No. 1A). When the coleoptiles are about 2.5 cm long, they are ready for use.

Cut twenty 8 mm segments of coleoptiles using a two-bladed cutter. The cutting should be performed under dim red light. Use only one segment from immediately below the 2 mm tip of each coleoptile. Carefully remove the leaf from each segment and string ten coleoptiles on each of two cotton threads using a suitable needle. Then stack the coleoptiles on a 1 cm square piece of plastic sheet held by a knot on the thread. Again, be sure to carry out this procedure under dim red light. Measure the total length of both of the ten segment stacks (zero time length) using a binocular miscroscope. Immediately place one stack in buffer and the other stack in buffer with indoleacetic acid 5×10^{-6}M. Measure the total length of each stack of coleoptiles every half hour for two hours in dim light. During the two-hour period, place the petri dishes in a drawer or wrap them in aluminum foil.

An alternate procedure involves one string of strung coleoptiles placed first in buffer for twenty to thirty minutes and then sequentially for twenty to thirty minutes each in 10^{-7}M indoleacetic acid, 10^{-6}M indoleacetic acid, and 10^{-5}M indoleacetic acid. The coleoptiles should be blotted with tissue before measuring change.

Results/ Conclusions

Construct a table comparing the growth of the two coleoptiles stacks. Why was the tip of the coleoptile removed before use? What is the effect of the indoleacetic acid (auxin)? How does auxin function in plant growth? How might it function on the molecular level?

Acknowledgment   This exercise was suggested by Dr. Michael Evans, The Ohio State University.

References   Bonner, J. 1934. The relation of hydrogen ions to the growth rate of the *Avena* coleoptile. *Protoplasma* 21: 406–423.

Cleland, R. E. 1982. The mechanism of auxin-induced proton efflux. In Waring, P. F. (ed.), *Plant Growth Substances*. London: Academic Press, pp. 23–31.

Cohen, J. D. and K. D. Nadler. 1976. Calcium requirement for indoleacetic acid-induced acidification by *Avena* coleoptiles. *Plant Physiol.* 57: 347–350.

Evans, M. L. 1974. Rapid responses to plant hormones. *Ann. Rev. Plant Physiol.* 25: 195–223.

Evans, M. L. 1976. A new sensitive auxanometer. *Plant Physiol.* 58: 598–601.

Rayle, D. L. 1973. Auxin-induced hydrogen-ion secretion in *Avena* coleoptiles and its implications. *Planta* 114: 68–73.

EXERCISE 53

Proton pumping and straight and gravity-directed growth

Introduction

One of the actions of auxin in cellular elongation appears to reside in its ability to induce a decrease in pH in the vicinity of the cell wall. In line with this observation, Rayle and Cleland (see the references that follow) proposed that auxin-induced acidification is the mechanism by which cell wall deformation takes place. In the presence of auxin, the pH at a site in the cell wall becomes acidic; this may in turn activate cell wall loosening enzymes or the protons may act directly in the wall cross-linkages and cause breaks between the non-covalent bands of the structural chemicals.

It has been suggested that auxin induces a decrease in the vicinity of the cell wall by activating a membrane-bound H^+ ion pump present in the plasmalemma. However, there is little direct evidence that supports this idea.

A very simple method of demonstrating the action of auxin with respect to pH changes involves the use of corn seedlings. Observations are taken during straight root growth and/or geotropic curvature while the roots are imbedded in agar containing bromcresol purple. The bromcresol purple will change color as the pH changes (see the results/conclusions section) and will indicate acid efflux from the roots into the agar-dye medium.

Purpose

To illustrate acid efflux from roots during growth and tropistic curvature.

Materials

corn seedlings, *Zea mays* L.
(enough so that each student or student group will have four or six straight roots, ready for use three days after germination)

medium (4 mm thick in 10 cm petri dishes), freshly prepared and containing 0.6% agar with 0.71 mM bromcresol purple indication dye

India ink and applicator
plastic boxes with lids (one per
    student group), large enough to

hold 10 cm petri dishes vertically and lined with wetted filter
paper to maintain humidity

Procedure

A.   Acid efflux from roots of corn seedlings

Mark the roots of three-day-old corn seedlings with India ink at millimeter intervals. Take one seedling and carefully press the length of the primary root, up to one-half the circumference, into the agar in the dish supplied by the instructor. The root side of the seedling should be pressed into the medium. Four seedlings should be used per group. Cover the dishes with lids and hold them vertically (and the root vertically) in the plastic box. After adding wetted filter paper to keep the humidity high, close the box. At five-minute intervals, make observations of color changes of the indicator in the medium and associate them with regions (root cap cell enlargement) of the root. Sketch the root and the color of the medium at each five-minute interval. Stipling, lines, crosses, and so forth may be used to indicate different colors. See Figure 53–1.

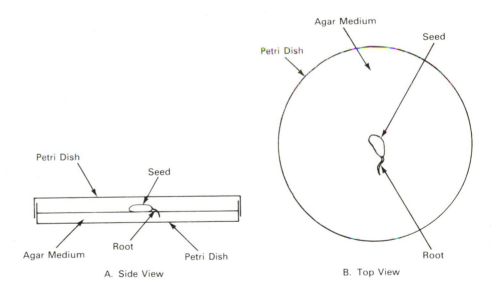

FIGURE 53–1   Proton pumping.

B.   Acid efflux and curvature of corn root

Place the seedling primary root, up to one-half the circumference, into the surface of the agar in the petri dish. The dish should be held vertically in the plastic box with the root oriented vertically as well. Four dishes should be set up for each group. Maintain humidity in the plastic box by inserting wet filter paper in the box as in Part A. Observe at twenty-minute intervals the color changes associated with the positive geotropic curvature.

Results/ Conclusions

When you discuss your results, which should include drawings of the root and the color associated with each portion and side of the root, keep in mind that the indicator dye bromcresol purple changes color in the range of pH 3.5 to 10. From 3.5 to 4.8 it is yellow, from 4.8 to 5.5 it is orange, from 5.5 to 6.4 it is red, and from 6.4 to 10 it becomes dark red and then violet.

In your discussion of the positive geotropic response exhibited by the roots, relate the observed color change to the current ideas of gravity-directed growth. Include a summary of the Cholodny-Went (differential-growth) hypothesis, and from your reading suggest alternatives to this hypothesis.

References

Cleland, R. E. 1976. Kinetics of hormone-induced H^+ excretion. *Plant Physiol.* 58: 210–213.

Cleland, R. E. 1982. The mechanism of auxin-induced proton efflux. In Waring, P. F. (ed.), *Plant Growth Substances*. London: Academic Press, pp. 23–31.

Cohen, J. D. and R. S. Bandurski. 1982. Chemistry and physiology of bound auxins. *Ann. Rev. Plant Physiol.* 33: 403–430.

Firn, R. D. and J. Digby. 1980. The establishment of tropic curvature in plants. *Ann Rev. Plant Physiol.* 31: 131–148.

Mulkey, T. J. and M. L. Evans. 1981. Geotropism in corn roots: Evidence for its mediation by differential acid efflux. *Science* 212: 70–71.

Mulkey, T. J., K. M. Kuzmanoff, and M. L. Evans. 1981. Correlations between proton-efflux patterns and growth patterns during geotropism and phototropism in maize and sunflower. *Planta* 152: 239–241.

Mulkey, T. J., K. M. Kuzmanoff, and M. L. Evans. 1981. The agar-dye method for visualizing acid efflux patterns during tropistic curvatures. *What's New In Plant Physiology* 12 (3): 9–12.

EXERCISE 54

Indoleacetic acid oxidase from pea epicotyls

Introduction

One means of inactivation or destruction of IAA in plants seems to reside in the activity of the IAA–oxidase enzyme system. In fact, the major disappearance of the phytohormone *in vivo* appears to be due to IAA–oxidase activity. IAA oxidase seems to be ubiquitous, having been isolated from numerous plant sources. The idea that the IAA–oxidase system is also a peroxidase seems to be supported by considerable current evidence; the enzyme system responsible for IAA destruction always shows conventional peroxidase activity.

Peroxidase activity is characterized by the oxidation of phenols with hydrogen peroxide (H_2O_2) as the electron acceptor accordingly:

$$H_2O_2 + \text{phenol (reduced)} \xrightarrow{\text{peroxidase}} \text{phenol (oxidized)} + 2H_2O$$

The major distinction between peroxidase and oxidase is that a peroxidase reaction does not require the addition of oxygen. In the IAA–oxidase reaction on IAA, the peroxidative enzyme system operates as an oxidase, as illustrated by the following:

IAA   3-methyleneoxindole

Although not illustrated, the reaction of some IAA–oxidase enzymes require the presence of Mn^{+2} and a phenolic factor such as 2,4-dichlorophenol. Further, electrophoresis studies on IAA oxidases from different plant sources show that IAA oxidases may exist as multiple enzyme forms. The exact nature of the different

237

forms as well as their significance in regulating the levels of IAA *in vivo* are not known.

Purpose

To extract the indoleacetic acid oxidase from the epicotyls of pea seedlings and to investigate the enzymatic destruction of indole-3-acetic acid (IAA) *in vitro*.

Materials

100 pea seedlings (*Pisum sativum* L. 'Alaska') grown in total darkness for seven to nine days (see Procedure, Part A)

100 ml phosphate-citrate buffer (equal volumes of 0.2M Na_2HPO_4 and 0.1M citric acid; adjusted to pH 5.6)

10 ml indoleacetic acid solution (200 μg/ml)

40 ml acetone

$FeCl_3$ solution (0.5M)

$HClO_4$ solution (35%, w/v)

6 Erlenmeyer flasks (25 ml)

6 test tubes

1 test tube rack

cheesecloth

centrifuge (refrigerated, if available)

6 colorimeter tubes or cuvettes

spectrophotometer (set as 540 nm, or Klett-Summerson colorimeter equipped with a No. 54 filter)

Procedure

A.   Plant material

Soak pea seeds (*Pisum sativum*, L. 'Alaska') in running tap water for four to six hours. Sow the soaked seeds in flats of vermiculite or sand and grow them in the dark for seven to nine days.

B.   Enzyme preparation

From the dark-grown seedlings, harvest approximately 32 g (fresh weight) of the etiolated epicotyls and homogenize them with a minimum amount of cold distilled water in a chilled homogenizer (Ten Broeck or other glass tissue grinder) or large mortar.

Pass the homogenate through a layer of cheesecloth and centrifuge the liquid at 10,000 × g for fifteen minutes. Collect the supernatant and add enough acetone so that the final acetone concentration is 40% by volume. The resulting precipitate contains the enzyme.

Centrifuge the above at 600 × g for fifteen minutes. Discard the supernatant and resuspend the pellet in 8 ml of buffer (pH 5.6). Each milliliter of this enzyme preparation represents 4 g fresh weight of tissue. This preparation may be frozen and stored (no longer than three weeks) or used immediately.

C. IAA–oxidase reaction

To perform a simple test for IAA oxidase activity add the stock solution to 25 ml Erlenmeyer flasks as outlined in Table 54–1.

TABLE 54–1   IAA–oxidase reaction mixture

Stock Solutions	Amount Added (ml) to Tube Number					
	1	2	3	4	5	6
Buffer (pH 5.6)	9.0	9.0	8.0	8.0	7.0	6.0
IAA (200 μg/ml)	0	1.0	1.0	1.0	1.0	1.0
Enzyme preparation	1.0	0	1.0 (Boiled)	1.0	2.0	3.0
Total volume	10.0	10.0	10.0	10.0	10.0	10.0

Do not add enzyme until the start of the reaction. In flask number 3, use enzyme heated for fifteen minutes in a boiling water bath. For the heated enzyme, adjust back to original volume with buffer if necessary and cool to room temperature before use.

Start the reaction by adding enzyme. Then mix the contents of each flask by gentle shaking. Allow the reaction mixtures to incubate at room temperature for fifty minutes.

D. Salkowski test and residual IAA determination

During the incubation time of the mixtures in Part C, number six test tubes corresonding to each reaction mixture. Then add 8 ml of Salkowski reagent (1 ml 0.5M FeCl$_3$ in 50 ml of 35% HClO$_4$ — *Caution: perchloric acid is caustic!*) to each test tube.

In order to stop the enzyme-catalyzed reaction at the end of fifty minutes, pipet 4 ml aliquots of each reaction mixture into the correspondingly labeled test tube containing the Salkowski reagent. Then carefully mix the contents. Allow the tubes to stand at room temperature for thirty minutes to ensure full color development.

After thirty minutes, pour the contents of each tube into a cuvette and determine the residual IAA on the basis of color development in a Klett-Summerson photoelectric colorimeter equipped with a No. 54 filter or with a Spectronic 20 spectrophotometer set at 540 nm. Tube No. 1 is to be used as a reagent blank.

Using the Klett units or optical density units obtained, calculate the amount of IAA destroyed by comparing the values obtained for the reaction mixture containing active enzyme (4, 5, and 6) with that of the reaction mixture containing inactivated enzyme (flask 3).

If desired, additional experimentation may be performed (using the basic steps presented) to study the effects of time, different substrate concentrations, and/or pH on the reaction rate. These experiments are left to the discretion of the instructor.

Results/
Conclusions

Record your results in Table 54–2. In concluding statements, interpret the results obtained and discuss the general procedures outlined for the preparation of the enzyme (for example, why were dark-grown seedlings used as the enzyme source?) and for the enzyme assay. What are the advantages and limitations of the Salkowski test for determining residual IAA? Also, present some of the current ideas pertaining to the nature of the enzyme in question and the possible mechanism of action involved in the enzymatically catalyzed destruction of IAA *in vivo* and *in vitro*.

TABLE 54–2   IAA–oxidase activity

Reaction Mixture* Flask Number	Optical Density or Klett Units	Residual IAA (µg)	IAA Destroyed (µg)
1			
2		200	
3			
4			
5			
6			

* Reaction mixtures consisted of the various components outlined in Table 54–1 and after the addition of enzyme were incubated for fifty minutes at room temperature.

Acknowledgment

This exercise was adapted for class use from the work of R. S. Rabin and R. M. Klein.

References

Galston, A., J. Bonner, and R. J. Baker. 1953. Flavoprotein and peroxidase as components of the indoleacetic acid oxidase system of peas. *Arch. Biochem. Biophys.* 42: 456–470.

Galston, A. W. and L. Y. Dalberg. 1954. The adaptive formation and physiological significance of indoleacetic acid oxidase. *Amer. J. Bot.* 41: 373–380.

Gordon, S. A. 1954. Occurrence, formation, and inactivation of auxin. *Ann. Rev. Plant Physiol.* 5: 341–378.

Gordon, S. A. and R. P. Weber. 1951. Colorimetric estimation of indoleacetic acid. *Plant Physiol.* 26: 192–195.

Larsen, P. 1951. Formation, occurrence, and inactivation of growth substances. *Ann. Rev. Plant Physiol.* 2: 169.

Lipetz, J. and A. W. Galston. 1959. Indoleacetic acid oxidase and peroxidase in normal and crown-gall tissue cultures of *Parthenocissus tricuspidata. Amer. J. Bot.* 46: 193–196.

Maclachlan, G. A. and E. R. Waygood. 1956. Kinetics of the enzymatically catalyzed oxidation of indoleacetic acid. *Can. J. Biochem. Physiol.* 34: 1233–1250.

Platt, R. S., Jr. and K. V. Thimann. 1956. Interference in Salkowski assay of indoleacetic acid. *Science* 123: 105–106.

Rabin, R. S. and R. M. Klein. 1957. Chlorogenic acid as a competitive inhibitor of indoleacetic acid oxidase. *Arch. Biochem. Biophys.* 70: 11–15.

Siegel, B. Z. and A. W. Galsoton. 1967. Indoleactic acid oxidase activity of apoperoxidase. *Science* 157: 1557–1559.

Wagenknecht, A. C. and R. H. Burris. 1950. Indoleacetic acid inactivating enzymes from bean roots and pea seedlings. *Arch. Biochem.* 25: 30–53.

Waygood, E. R., A. Oaks, and G. A. Maclachlan. 1956. On the mechanism of indoleacetic acid oxidation by wheat leaf enzymes. *Can. J. Bot.* 34: 54–59.

Yamazaki, I. and H. Souzu. 1960. The mechanism of indoleacetic acid oxidase reaction catalyzed by turnip peroxidase. *Arch. Biochem. Biophys.* 86: 294–391.

EXERCISE 55

Gibberellic acid and plant growth

The gibberellins belong to a large group of compounds called terpenoids and occur naturally in a large number of plants. The gibberellins consist of the ent-gibberellane skeleton (twenty carbons) or ent-20 norgibberellanes (nineteen carbons). They are distinguished also by the presence or absence of the lactone configuration (internal ester) in the A ring with substituents (particularly hydroxyl groups) about the ring structure. The structure of gibberellic acid (GA_3) is as follows:

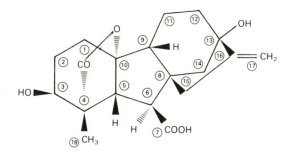

Gibberellic acid (GA_3)

Transport of the gibberellins is for the most part nonpolar in the phloem with flow patterns similar to those of carbohydrates and other organic substances. Gibberellins may also be translocated in the xylem due to lateral movement between the two vascular tissues. The actual mechanism(s) involved in the distribution of gibberellins within the plant is not known.

Gibberellins often have been compared in biological activity with the auxins. For example, they and IAA both promote cellular elongation, promote cambial activity, induce pathenocarpy, and stimulate nucleic acid and protein synthesis.

Applied gibberellins promote internodal elongation of dwarves and plants showing intermediate growth rates. It is interesting to note that, in contrast to the

242

auxins, gibberellins promote stem elongation of intact plants, not the elongation of stem segments. The following exercise illustrates the effect of gibberellic acid in the elongation of plants.

Purpose

To study the effect of gibberellic acid on plant growth.

Materials

20 pea seedlings (*Pisum sativum* L. 'Little Marvel'; see Procedure, Part A)

30 bean plants (*Phaseolus vulgaris*, L. 'Pinto'; five per pot; see Procedure, Part A)

gibberellic acid solution (100 mg/l), dissolved in 1 to 2 ml 95% ethanol and diluted to 1 liter with water

5 solutions of different GA_3 concentrations (10 ml of each solution):

— GA_3 10^{-1}M (0.01 × gram molecular weight of GA_3 dissolved in 1 ml ethanol and diluted to 100 ml with water)

— GA_3 10^{-2}M (1 ml 10^{1}M GA_3 diluted to 10 ml with water)

— GA_3 10^{-3}M (1 ml 10^{-2}M GA_3 diluted to 10 ml with water)

— GA_3 10^{-4}M (1 ml 10^{-3}M GA_3 diluted to 10 ml with water)

— GA_3 10^{-5}M (1 ml 10^{-4}M GA_3 diluted to 10 ml with water)

1000 ml distilled water containing 1 to 2 ml 95% ethanol

flat containing vermiculite

plastic sheets (for covering plants)

ruler

Procedure

A. Plant material

Both tall and dwarf varieties of bean, corn, or pea plants may be used according to procedures outlined below. The following techniques are presented primarily as a working guide to illustrate the effect of gibberellic acid on plant growth.

1. Dwarf pea seedlings (two weeks old): Sow two rows of dwarf pea seeds (*Pisum sativum* L. 'Little Marvel') in a flat containing vermiculite. Plant at least ten seeds per row. Cover the seeds with ½ in. of vermiculite and maintain the flat in the greenhouse for two weeks with occasional watering. Treat the two-week-old plants according to Part B.

2. Potted bean plants (two to three weeks old): Sow at least five bean seeds (*Phaseolus vulgaris* L. 'Pinto') in each of six pots containing soil. Cover the seeds with ½ in. of soil, water, and maintain in the greenhouse for two to three weeks. When the plants are ready for use, select two plants in each pot so that all plants selected among the six pots are uniform in height. Harvest those plants at soil level that are not to be used. Treat the potted plants (two per pot) according to the directions in Part C.

B. Gibberellic acid treatment and plant growth response

Cover one of the rows of the two-week-old pea plants with a protective plastic cover and thoroughly spray the other row with an aqueous solution of GA_3 (100 mg/l). After the plants have dried, cover the treated row and spray the first row with the distilled water containing the same amount of ethanol as was used to prepare the GA_3 solution.

Leave the plants in the greenhouse and record any difference in appearance over the next three to four weeks. At least three times a week, measure the height of the plants (from soil level), the length of internodes, length of leaves (from the base of the midrib to the leaf tip), and the blade width. At the end of each week, spray the plants as initially performed. Remember to calculate an average for each measurement of the control and treated plants.

Observations and measurements:

C. Effect of applications of gibberellic acid to stem tip

Label the potted bean plants (described in Part A) according to the following treatments:

Treatment 1: distilled water (control)

Treatment 2: GA_3 ($10^{-1}M$)

Treatment 3: GA_3 ($10^{-2}M$)

Treatment 4: GA_3 ($10^{-3}M$)

Treatment 5: GA_3 ($10^{-4}M$)

Using an eye dropper, deposit a drop of distilled water or the appropriate GA_3 solution on the growing tip of the two plants in each appropriately labeled pot. (Label the control or treated plant in each pot.) Carefully rinse the dropper with distilled water after each application or use separate droppers. Also, be sure to use a distilled water solution containing the same amount of ethanol used in the GA_3 solutions.

After making the initial applications, maintain the plants on a greenhouse bench and record any changes in appearance during the next three to four weeks. Measure the plants at least three times a week with respect to total height (from soil level), length of internodes, and length and width of leaves. After making the last measurements each week, repeat the application with the various test solutions.

Observations and measurements:

Results/ Conclusions

Present a table for each part of the experiment in which the averages of the measurements made on the control and treated plants are illustrated. Indicate the major growth response exhibited by the treated plants that can be attributed to the gibberellic acid treatment. Is GA_3 translocated from the site of application to other areas on the treated plants? In addition to stimulating cellular elongation, what other plant responses do the gibberellins influence?

In additional remarks, discuss the possible site of GA_3 synthesis in plants and the chemical structure of gibberellins.

Does dwarfism of plants appear to be due to the genetically controlled absence of gibberellins, the presence of specific inhibitors, or to some form of repressed cellular sensitivity to the gibberellins? Explain.

References

Brian, P. W., G. W. Elson, H. G. Hemming, and M. Radley. 1954. The plant growth promoting properties of gibberellic acid, a metabolic product of the fungus *Gibberella fujikuroi. J. Sci. Food Agric.* 5: 602–612.

Brian, P. W. and H. G. Hemming. 1955. The effect of gibberellic acid on shoot growth of pea seedlings. *Physiol. Plantarum* 8: 669–681.

Bukovac, M. J. and S. H. Wittwer. 1957. Gibberellin and higher plants. II. Induction of flowering in biennials. *Quar. Bull. Mich. Agric. Exp. Sta.* 39(4): 650–660.

Chin, R. Y. and J. A. Lockhart. 1965. Translocation of applied gibberellin in bean seedlings. *Amer. J. Bot.* 52: 828–833.

Crozier, A. (ed.). 1983. *The Biochemistry and Physiology of Gibberellins.* New York: Praeger Scientific.

Graebe, J. E., D. T. Dennis, C. D. Upper, and C. A. West. 1965. Biosynthesis of gibberellins. *J. Biol. Chem.* 240: 1847–1854.

Hedden, P., J. MacMillan, and B. O. Phinney. 1978. The metabolism of the gibberellins. *Ann. Rev. Plant Physiol.* 29: 149–192.

Jacobs, W. 1979. *Plant Hormones and Plant Development.* New York: Cambridge University Press.

Jones, R. L. 1973. Gibberellins: Their physiological role. *Ann. Rev. Plant Physiol.* 24: 571–598.

Jones, R. L. and J. MacMillan. 1984. Gibberellins. In Wilkins, M. B. (ed.), *Advanced Plant Physiology.* London: Pitman Publishing, Limited, pp. 21–52.

Moore, T. 1979. *Biochemistry and Physiology of Plant Hormones.* New York: Springer-Verlag.

Paleg, L. G. 1965. Physiological effects of gibberellins. *Ann. Rev. Plant Physiol.* 16: 291–322.

The effect of gibberellic acid on the production of reducing sugars: the basis for a bioassay

Introduction

Growth of the embryo during germination of grains depends on the enzymatic breakdown of the stored starch to simple sugars, followed by the translocation of the sugars to the embryo where they provide an energy source for growth. Gibberellins play an important role in the mobilization of carbohydrate reserves in cereal grains.

Until the late 1950s, it was believed that the endosperm played only a passive role in germination and that the embryo provided the enzymes for the mobilization of endosperm starch reserves. We now know that the aleurone layer of the endosperm is sensitive to gibberellins, and that when the phytohormone is released from the embryo during germination it stimulates the cells of the aleurone to release hydrolytic enzymes for the digestion of the endosperm starch. The enzymes released from the aleurone cells are α-amylase, ribonuclease, β-1,3-gluconase, protease, β-amylase and possibly other unidentified hydrolases. Also, it appears that the enzymes, with the exception of β-amylase, are synthesized *de novo* by the GA-stimulated aleurone cells. Further, when the starch enzymes (α- and β-amylase) are released into the endosperm, the starch is converted to simple sugars, which are used as an energy source for embryo growth.

As a result of the action of gibberellins on the mobilization of starch reserves in barley, this system has been used widely as a bioassay for gibberellins. In this assay, half-grains of barley (the half not containing the embryo) are incubated in a solution of gibberellic acid (GA_3) or unknown extract thought to contain gibberellins, for a prescribed time period. When gibberellins are present, they will stimulate amylase activity (synthesis and release from the aleurone cells) with the disappearance of starch and the buildup of reducing sugars. The amount of sugars is proportional to the concentration of gibberellins present. The following exercise illustrates the action of gibberellic acid on the barley endosperm system.

Purpose

To study the effect of gibberellic acid on anylase activity in barley endosperm as indirectly measured by the detection of reducing sugars.

Materials

60 barley grains (*Hordeum vulgare* L. 'Naked Blanco Mariout') soaked in H_2O overnight (see Procedure, Part A)

100 ml hypochlorite solution (1%) or Clorox solution (1%, v/v)

gibberellic acid stock solution (10 mg/100 ml H_2O)

5 separate gibberellic acid test solutions (see Procedure, Part B)

Somogyi's reagent (see Appendix II)

Nelson's arsenomolybdate reagent

(see Appendix II)

12 mg streptomycin sulfate (solid)

1 g Rexyn 101 (H^+) resin

spectrophotometer (set at 540 nm, or a Klett colorimeter equipped with a No. 54 filter)

7 colorimeter tubes (or cuvettes)

13 test tubes

5 volumetric flasks (100 ml)

pipets

water bath (boiling)

Procedure

A. Barley seed preparation

Soak approximately sixty barley seeds (*Hordeum vulgare* L. 'Naked Blanco Mariout') in 50% sulfuric acid (v/v) for one hour. Decant the acid and wash the seeds overnight in running tap water at a temperature of about 30°C.

B. Gibberellic acid solutions

Starting with the stock solution of gibberellic acid (10 mg/100 ml H_2O), prepare the various test solutions (numbered 1 through 5) by serial dilution according to the following:

Solution 1: 10^{-6} g GA_3/ml H_2O (dilute 1 ml of stock solution to 100 ml with water)

Solution 2: 10^{-7} g GA_3/ml H_2O (dilute 10 ml of solution 1 to 100 ml with water)

Solution 3: 10^{-8} g GA_3/ml H_2O (dilute 10 ml of solution 2 to 100 ml with water)

Solution 4: 10^{-9} g GA_3/ml H_2O (dilute 10 ml of solution 3 to 100 ml with water)

Solution 5: 10^{-10} g GA_3/ml H_2O (dilute 10 ml of solution 4 to 100 ml with water)

Pipet 10 ml of each test solution into separate flasks and dissolve 2 mg of streptomycin sulfate in each 10 ml aliquot. Then pipet 1 ml of each solution into a separate test tube and number the tubes 1 through 5. Into another tube (No. 6), pipet 1 ml of water (also containing 2 ml streptomycin sulfate per 10 ml).

C.   Gibberellic acid treatment of the seeds

Select fifty water-soaked seeds of the same size and rinse them briefly in 1% hypochlorite solution. Then wash them thoroughly with distilled water to remove the residual hypochlorite. Divide the seeds into six groups and cut each seed transversely in half. Discard the embryo half and save the endosperm half.

From the half-seeds, select four of uniform size from each group and place them into the appropriately labeled test tubes (four half seeds per tube) containing the 1 ml portions of 10^{-6}, 10^{-7}, 10^{-8}, 10^{-9}, 10^{-10} g GA_3/ml and the water control (tube No. 6). Stopper the tubes, place them on a shaker, and incubate them at room temperature for twenty-four hours. The shaker is not absolutely necessary, if one is not available.

At the end of the twenty-four-hour period, add 4 ml of distilled water to each test tube and store in a freezer until the next laboratory period. At the next laboratory period, thaw the tubes and to each add approximately 150 mg of solid Rexyn 101 (H^+) resin. Stopper the tubes tightly and shake the solutions thoroughly. Allow the resin to settle and then test a 2 ml aliquot of each solution for the presence of reducing sugars.

D.   Detection of reducing sugars

To six test tubes containing 2 ml Somogyi's reagent, add a 2 ml aliquot of the solution surrounding the seeds. Also, prepare a tube containing 2 ml of reducing sugar reagent and 2 ml distilled water (reagent blank).

Cover the test tubes with marbles and heat the reaction mixture in a boiling water bath for *exactly* ten minutes. Cool the tubes in a cold water bath for five minutes. Then add 2 ml of Nelson's arsenomolybdate reagent. After mixing the solutions, transfer the contents of each tube to a cuvette or colorimeter tube and measure the absorbency in a colorimeter equipped with a No. 54 filter or in a spectrophotometer set at 540 nm. Save the incubation mixtures containing the seeds since it may be necessary to repeat the color reaction.

Absorbency measurements for each solution:

Results/
Conclusions

Present your results in the form of a graph in which the optical density values or Klett units obtained for each sample are plotted as the ordinates against the respective concentration of gibberellic acid on the abscissa. How might the absolute amounts of reducing sugars in each solution be quantitatively determined?

Would the technique outlined here be adequate as a means to assay for gibberellins present in various plant extracts?

From your reading and the results obtained, present some ideas as to the action of gibberellic acid on the barley endosperm tissue. Also consider whether or not gibberellin, amylase activity, and the presence of reducing sugars in endosperm tissue are necessary for barley seed germination under normal conditions.

Acknowledgment

The general outline for this exercise was taken from the work of P. B. Nicholls and L. G. Paleg and adapted for class use by C. J. Pollard, Michigan State University, East Lansing.

References

Bewley, J. D. and M. Black. 1978. *Physiology and Biochemistry of Seeds.* I. *Development, Germination and Growth.* Berlin: Springer-Verlag.

Chrispeels, M. J. and J. E. Varner. 1966. Inhibition of gibberellic acid induced formation of α-amylase by abscission. II. *Nature* 212: 1066–1067.

Chrispeels, M. J. and J. E. Varner. 1967. Hormonal control of enzyme synthesis: On the mode of action of gibberellic acid and abscission in aleurone layers of barley. *Plant Physiol.* 42: 1008–1016.

Jones, R. L. 1973. Gibberellins: Their physiological role. *Ann. Rev. Plant Physiol.* 24: 571–598.

Krisknamoorthy, H. N. (ed.). 1975. *Gibberellins and Plant Growth.* New York: John Wiley and Sons.

Nicholls, P. B. and L. G. Paleg. 1963. A barley endosperm bioassay for gibberellins. *Nature* 199: 823–24.

Gibberellic acid and chlorophyll retention in *Taraxacum* leaf disks: the basis for a bioassay

Introduction

The stimulation of chlorophyll retention by gibberellins represents another form of biological activity exhibited by these compounds. This effect of gibberellins has also been employed in the past to study their action and as a bioassay. When leaf disks or leaf sections (*Rumex* leaves respond best) are incubated in a solution of GA, they will retain chlorophyll longer than the controls incubated in water or buffer. The assay period takes about four to five days for optimum comparison of the treated disks with the controls.

Another growth regulator, kinetin (representative of the cytokinins) will also delay chlorophyll loss from detached leaves and leaf disks or sections. In fact, in most cases the cytokinins seem to be much more effective than the gibberellins in retarding chlorophyll loss. Nevertheless, the mechanism of action of the gibberellins or cytokinins with respect to chlorophyll retention is not known. It is quite likely, however, that their action involving chlorophyll is somehow related to the maintenance of chloroplasts. The auxins in general do not appear to delay chlorophyll degradation and loss in detached leaves or leaf sections.

Purpose

To investigate the effect of gibberellic acid on chlorophyll retention in *Taraxacum* leaf disks, the basis for a rapid bioassay for gibberellin. Kinetin and indoleacetic acid are also used to demonstrate the specificity of the assay.

Materials

Taraxacum officinale Weber (potted plants or plants growing out of doors)

3 separate gibberellin solutions (mg/l: 0.05, 0.5, and 5)

3 separate kinetin solutions (mg/l: 0.05, 0.5, and 5)

3 separate indoleacetic acid solutions (mg/l: 0.05, 0.5, and 5)

100 ml acetone (80%, v/v)

container lined with water-moistened filter paper

10 petri dishes, each containing a pad of Whatman No. 1 filter

paper

paper toweling

cork borer (internal diameter about 1 cm)

polyethylene container or plastic bags

mortar and pestle

Buchner funnel with filter paper pads

sidearm suction flask

cuvettes

spectrophotometer

Procedure

A. Plant material

Potted plants of *Taraxacum officinale* Weber growing under natural light conditions in the greenhouse or plants collected from out of doors may be used.

B. Preparation of leaf disks

Detach mature and fully exposed leaves from *Taraxacum* plants and place them on several layers of moistened paper toweling. If young leaves are used, they should be incubated with their petioles in water for one to three days until they are uniformly pale green in color prior to being used as a source of disks. Using a cork borer, punch out 110 disks of a convenient diameter (about 1 cm) from the interveinal portion of the leaves. As soon as each disk is cut, place it in a container lined with water-moistened filter papers.

C. Effect of gibberellic acid, kinetin, and indoleactic acid on chlorophyll retention in leaf disks from *Taraxacum*

Prepare ten appropriately labeled petri dishes containing a pad of Whatman No. 1 filter paper (three dishes for each treatment and one for distilled water). Moisten the pad of filter paper in each dish with the corresponding test solution indicated in Table 57–1. (Use about 1 ml of test solution for filter paper pad of 7 cm diameter.) The leaf disks will become waterlogged and will blacken if there is too much solution on the filter paper.

TABLE 57–1   Effect of three growth substances upon chlorophyll retention in leaf
disks of *Taraxacum officinale*

Sample	Treatment (mg/l)	Absorbency (652 nm) of Disk Extracts		
		GA$_3$	K	IAA
0 days*	—			
4 days†	0			
	0.05			
	0.5			
	5.0			

* Use an aliquot of the disks to determine total chlorophyll at the beginning of the experiment.

† Use one dish for the water control and enter the same reading under each of the columns
where indicated.

Transfer ten leaf disks selected at random to each petri dish. Cover the
dishes and wrap with moistened paper toweling.

Place the dishes in a polyethylene container or plastic bag to maintain high
humidity and store in darkness at room temperature for four days. Determine the
total chlorophyll content according to the method of extraction and spectropho-
tometry outlined in Exercise 29. Since the amount of chlorophyll determined in
this sample and the experimental disks will be compared on a relative basis, there
is no need to calculate the absolute amounts of chlorophyll. Optical density readings of 80% acetone extracts at 652 nm will be sufficient to determine differences
in chlorophyll content.

After four days, determine the chlorophyll content of the total number of
leaf disks in each treatment. Record the optical denisty readings taken at 652 nm
for each treatment in Table 57–1.

**Results/
Conclusions**

Construct a bar graph with the optical density readings of the chlorophyll extracts
plotted against the different treatments. Interpret the results on the basis of the
effect of gibberellic acid upon the retention of chlorophyll in leaf disks of *Taraxacum*. In concluding remarks, indicate some of the other general biological responses attributed to the action of gibberellins. Does this assay seem to be specific
for gibberellins?

Acknowledgment

This exercise was adapted from the work of R. A. Fletcher and D. J. Osborne.

References

Crozier, A. (ed.). 1983. *The Biochemistry and Physiology of Gibberellins*. New
York: Praeger Scientific.

Fletcher, R. A. and D. J. Osborne. 1965. Regulation of protein and nucleic acid
synthesis by gibberellin during leaf senescence. *Nature* 207: 1176–1177.

Fletcher, R. A. and D. J. Osborne. 1966. Gibberellin as a regulator of protein and ribonucleic acid synthesis during senescence in leaf cells of *Taraxacum officinale. Can. J. Bot.* 44: 739–745.

Jones, R. L. 1973. Gibberellins: Their physiological role. *Ann. Rev. Plant Physiol.* 24: 571–598.

Osborne, D. J. and D. R. McCalla. 1961. Rapid bioassay for kinetin and kinins using senescing leaf tissue. *Plant Physiol.* 36: 219–221.

Paleg, L. G. 1965. Physiological effects of gibberellins. *Ann. Rev. Plant Physiol.* 16: 291–332.

(Also see the references following Exercises 55 and 56.)

EXERCISE 58

The culture of carnation meristems

Introduction

The early work on the culture of plant organs and tissues provided the framework for detecting phytohormones in plant extracts, a broader understanding of cellular division and enlargement, and further knowledge relating to cellular differentiation and morphogenesis. It was not until the mid-1960s, however, that many of the commercial implications of tissue and organ culture techniques were recognized. Today there is much emphasis placed on the propagation of plants through tissue-culture techniques. One means to produce whole plants from plant tissue involves the isolation and culture of meristematic tissue. Culturing meristems has considerable advantages, among them the simplicity and low cost of producing uniform, reproducible, and high-quality plants. Meristem culture with subsequent development of plants also allows growers to produce a continuous supply of plants without using expensive space as a stock-holding facility. Virus-free plants, such as geranium, are now being produced in great quantity because of meristem tissue-culture techniques and are shipped to world markets. The following exercise will illustrate the techniques in the culture of carnation meristems.

Purpose

To culture aseptically the meristem of carnations.

Materials

test tubes with autoclavable caps (four per student)
test tube rack
filter paper (Whatman No. 1)
liquid medium (see below)
autoclaved petri dishes (four per student)
pipets (10 ml), volumetric
scissors (small)
carnations, vegetative shoots, 6 to 7 in. long (five or six per student, or enough so that at least four

"tips" are obtained per student)
nasal forceps (small)
forceps (small)
scalpel, handle with No. 11 blade
alcohol lamp
flask, with 95% ethanol for flaming forceps
transfer hood (laminar flow hood)
constant temperature or growth chamber (set at 24°C with continuous light)

Procedure

A. **Culture medium**

The culture medium, Sutton's 1971 in de Fossard 1976 (see the references at the end of this exercise), is outlined in Table 58–1.

TABLE 58–1   A preparation of culture medium

Compound	Amount per Liter	
	Mass	Mole
$CA[NO_3]_2 \cdot 4H_2O$	1000 mg	4.23 mmol
KNO_3	250 mg	2.47 mmol
$MgSO_4 \cdot 7H_2O$	250 mg	1.01 mmol
KH_2PO_4	250 mg	1.84 mmol
$MnSO_4$	1 mg	6.62 μmol
KI	0.25 mg	1.51 μmol
$ZnSO_4 \cdot 7H_2O$	0.05 mg	0.174 μmol
$CuSO_4 \cdot 5H_2O$	0.025 mg	0.100 μmol
H_3BO_3	0.025 mg	0.404 μmol
$CoCl_2 \cdot 6H_2O$	0.025 mg	0.105 μmol
$NiCl_2 \cdot 6H_2O$	0.025 mg	0.105 μmol
$H_2SO_4[36N]$	0.5 μ	0.5 μmol
$FeSO_4 \cdot 7H_2O$	21.2 mg	76 μmol
$Na_2 \cdot EDTA$	28.3 mg	76 μmol
Glucose	40 g	222 μmol
NAA (Napthalene acetic acid)	1 mg	5.37 μmol
Thiamine·HCl	1 mg	2.96 μmol

Note: Adjust to pH 4.4 to 5.0 with diluted NaOH and HCl.

Add 5 ml of medium to each of four test tubes. Then cut four strips (2 × 14 cm) of filter paper and place one in each of the test tubes with medium, as in Figure 58–1. Using autoclavable caps or plugging cotton, cap the four test tubes. Autoclave the tubes for fifteen minutes at 15 lbs. pressure, 121°C. Allow the test tubes to cool to room temperature before use.

B. **Plant material**

The work area should be washed with 10% hypochlorite, and your hands should be washed with soap. Carefully strip back the outer leaves of the carnation shoot with your fingers. Then, using the small forceps, carefully strip the inner (4 mm) leaves from the shoot. Continue this until the stem tip is exposed.

After the exposure of the tip, cut it off in an autoclaved petri dish and replace the cover of the dish. A total of four stem tips should be cut and stored in autoclaved petri dishes. Using the flamed and cooled nasal forceps, lightly grasp one of the tips close to the cut end and transfer to the filter paper bridge in the

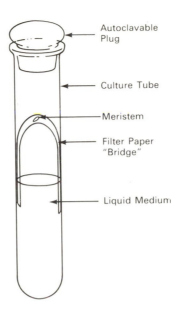

- Autoclavable Plug
- Culture Tube
- Meristem
- Filter Paper "Bridge"
- Liquid Medium

FIGURE 58-1   Meristem culture.

culture tube. Alternatively, the tip, still on the scalpel, may be directly placed on the filter paper bridge. The tubes should be held at an angle when open to reduce contamination. Place the remaining tips aseptically on the bridges in the other culture tubes. The "planted" culture tubes should be kept at 24°C in continuous light. At the discretion of the instructor, half of the class could place their cultures in a "cold box" at 4°C for two days before placing in the culture room. When roots are visible, transfer aseptically to new tubes containing the same medium with no NAA and with 20 g/l of glucose.

Results/ Conclusions

Make observations of the cultures on a weekly basis. What is the importance of meristem culture? What effect is produced by the auxin used (NAA)? What is the effect of auxins such as NAA on rooting and on root growth?

References

de Fossard, R. A. 1976. *Tissue Culture for Plant Propagators*. Armidale, N.S.W., Australia. The University of New England Printery, pp. 199–202.

Hu, C. Y. and P. J. Wang. 1983. Meristem, shoot tips, and bud cultures. In Evans, D. A., W. R. Sharp, P. V. Ammirato, and Y. Yamanda (eds.), *Handbook of Plant Cell Culture*, Vol. 1. New York: Macmillan Publishing Company, pp. 177–227.

Kartha, K. K. 1981. Meristem culture and cryopreservation-methods and application. In Thorpe, T. A. (ed.), *Plant Tissue Culture*. New York: Academic Press, pp. 181–211.

Murashige, T. 1974. Plant propagation through tissue cultures. *Ann. Rev. Plant Physiol.* 25: 135–166.

Pennazio, S. 1975. Effect of adenine and kinetin on development of carnation meristem tips cultured *in vitro. J. Hort. Sci.* 50: 161–164.

Roest, S. and G. S. Bokelmann. 1981. Vegetative propagation of carnation *in vitro* through multiple shoot development. *Sci. Hort.* 14: 357–366.

Seibert, M. 1976. Shoot initiation from carnation shoot apices frozen to −196°C. *Science* 119: 1178–1179.

Tissue culture techniques

Introduction

During the late 1950s and early 1960s, tissue cultures of soybean callus, derived from cotyledons, were developed and used by Miller as a bioassay to detect naturally occurring cytokinins. Normally, the cotyledons of *Glycine max,* cultivar 'Acme' is used for establishing the callus in culture. However, other varieties can be used. In general, the techniques used to establish the callus involve the germination and growth of soybean seedlings under sterile conditions. Aseptic tissue may be obtained by growing the seedlings in culture flasks in a nutrient medium with subsequent excision and transfer of cotyledon pieces to culture medium favoring callus formation. Another method involves the germination of seeds (after surface sterilization) in petri dishes or other suitable containers of vermiculite. The cotyledons may be surface-sterilized and planted as pieces in tissue-culture flasks in the appropriate culture medium. Pieces of cotyledon that do not appear to be contaminated with microorganisms may be selected and subcultured with subsequent callus formation. Once the callus tissue is established, it may be used as a bioassay or for studies on cell division and enlargement. The effects of kinetin or other cytokinins may be evaluated.

Purpose

To perform the basic techniques used for establishing cultures of callus tissue derived from various plant parts, and specifically to grow sufficient quantities of callus tissue from soybean (*Glycine max*) cotyledons to demonstrate its usefulness as a cytokinin bioassay in a subsequent experiment.

Materials

20 soybean seeds (*Glycine max* L. 'Acme'; other cultivars may also be used — see Appendix II); or young plants of bean (*Phaseolus vulgaris* L.), sunflower (*Helianthus annuus* L.), or tobacco (*Nicotiana tabacum* L.)

stock solutions (see Table 59–1)

60 g sucrose

20 g agar (powdered or flake)

200 ml Clorox solution (10%, v/v)

ethanol (95%)

cotton (nonabsorbent)

Erlenmeyer flasks (two 2 l; four 500 ml; twenty 125 ml)

2 volumetric flasks (1 liter)

10 pipets (5 or 10 ml, graduated)

8 beakers (400 ml), each containing 150 ml sterile distilled water and covered with aluminum foil lids

5–10 sterile petri dishes wrapped in paper bags (see Procedure)

pH meter

autoclave

transfer hood or room (not absolutely necessary)

Bunsen burner

balance (analytical)

forceps (length: 250 mm)

scalpel

Procedure

Before completing Part A of the procedure, read Part C and decide which method is to be employed to obtain sterile tissue for callus formation in culture. If the first method in Part C is to be followed, then complete Part A first. If the second method of Part C is preferred, omit Part A and go on to Part B.

A. Basal medium for growth of seedlings under aseptic conditions

Add approximately 500 ml of distilled water to a 2 l Erlenmeyer flask. Then add 5 ml of each of the stock solutions numbered 1 through 8 as indicated in Table 59–1. Do not add solutions 11 or 12.

Dissolve 30 g of sucrose in the solution and add distilled water until the volume is approximately 950 ml. Adjust the pH to 5.8 with sodium hydroxide and then the final volume to 1 l with water.

To the above completed solution, add 10 g agar and subsequently autoclave for three to five minutes or heat until the agar is melted. Be sure the agar is completely melted and mixed.

Pour 250 ml of the melted agar medium into each of four 500 ml Erlenmeyer flasks. Stopper the flasks with nonabsorbent cotton plugs and sterilize by autoclaving for fifteen minutes at 121°C and 15 lb./sq. in. pressure. It will also be convenient at this time to sterilize four 400 ml beakers, each covered with aluminum foil and containing about 150 ml of distilled water (to be used in Part C).

After sterilization, allow the flasks to stand at room temperature until the medium has solidified. The medium may be used immediately after hardening or stored in a refrigerator for several days before using.

TABLE 59–1 Stock solutions for callus tissue cultures*

Stock Solution Number	Component	Amount (g)	Final Aqueous Volume (ml)
1	KH_2PO_4	30.000	500
2	KNO_3	100.000	
	$Ca(NO_3)_2 \cdot 4H_2O$	50.000	500
3	$MgSO_4 \cdot 7H_2O$	7.150	
	KCl	6.500	500
4	$MnSO_4 \cdot H_2O$	0.492	
	$ZnSO_4 . 7H_2O$	0.380	500
5	H_3BO_3	0.160	
	KI	0.080	500
6	NH_4NO_3	100.000	500
7	EDTA $Na_2 \cdot 2H_2O$	1.340	
	$FeSO_4 \cdot 7H_2O$	0.990	500
8	$Cu(NO_3)_2 \cdot 3H_2O$	0.035	500
9	$(NH_4)_6Mo_7O_{24} \cdot 4H_2O$	0.010	500
10	Thiamin · HCl	0.080	500
	Nicotinic acid	0.200	
	Pyridoxine · HCl	0.080	500
	Myo-inositol	10.000	
11	α NAA	0.040	100
12	Kinetin	0.020	1000

* The stock solutions as given are generally prepared by investigators performing numerous experiments. If the solutions are to be only for class use, the amount of the components and final solution volumes should be scaled down accordingly.

B. Growth medium for callus tissue

Prepare a liter of medium as in Part A, but this time add 5 ml of the stock solutions 1 through 11 and 25 ml of solution 12 (kinetin). Dissolve 30 g of sucrose and bring the solution volume to about 950 ml with distilled water. Adjust the pH to 5.8 and then the final volume to 1 l. Add agar (10 g/l) and, after melting, pour 50 ml of the medium into each of twenty 125 ml Erlenmeyer flasks. Now wrap five petri dishes in a paper bag and, if Part A was omitted, fill four 400 ml beakers with 150 ml of distilled water and cover with aluminum foil. Then sterilize the flasks containing the medium, the petri dishes, and the beakers by autoclaving for fifteen minutes at 121°C and 15 lb./sq. in. pressure.

C. Preparation of sterile tissue for culture

To obtain tissue for culture, either one of the following methods may be used:

1. Sterilize twenty seeds (preferably soybean, *Glycine max* L. 'Acme'). Any one of the other cultivars listed in the Appendix may be used by soaking them in 10% Clorox solution for five to ten minutes.

Remove the residual Clorox by transferring the seeds with sterile forceps to a beaker of sterile distilled water. (In all of the following manipulations, use standard aseptic techniques and sterilize the forceps by dipping in 95% ethanol and flaming before each operation.) Four beakers of sterile water should have been prepared according to the directions in Part A or B.

After all of the seeds are transferred to the water, cover the beaker with the sterile aluminum foil lid and allow the seeds to stand for three minutes. Then transfer the seeds to a second beaker of water and so on until the seeds have been rinsed a total of four times.

Using suitable techniques (see Figure 59–1), transfer five water-rinsed seeds to each of the 500 ml Erlenmeyer flasks containing medium (prepared beforehand as directed in Part A). Maintain the seeds at room temperature and under constant fluorescent lighting. When ready (in about two weeks), the cotyledons may be excised and aseptically transferred (use forceps) to sterile petri dishes as indicated in Part D.

FIGURE 59–1   Equipment for tissue culture techniques.

2. Seedlings of soybean, bean, or sunflower may be grown in flats of vermiculite or sand in the greenhouse. If this is the case, the tissue should be surface-sterilized for three to ten minutes in 10% Clorox and then rinsed four times in sterile distilled water (previously prepared in Part B). All operations should be performed according to standard aseptic technique. Forceps should be used for the transfers and should be dipped in ethanol and flamed before each operation (see Figure 59–1). After the final water rinse, continue on to Part D.

D. Tissue cutting and culture

A transfer hood or room is not essential if careful aseptic technique is used. Remember to sterilize all implements before cutting and transferring the tissue.

Transfer suitably prepared tissue (from Part C) to a sterile petri dish. Quickly cut the tissue into blocks (approximately 4 × 4 × 2 cm) and place one block on the callus growth medium of each 125 ml flask. After each planting, stopper the flask. When the planting is completed, place the flasks in a growth chamber (25 to 27°C and constant light with an intensity of approximately 40 ft.-c.). If a growth chamber is not available, the tissues may be maintained in the laboratory under constant fluorescent lighting. For the next three weeks, observe the cultures and discard those that are contaminated.

After three weeks' growth, subculture the wound callus by aseptically cutting it into pieces of approximately 10 mg and transferring the pieces to fresh callus growth medium (three pieces per flask). Also, make a thin freehand section from a piece of callus, mount on a microscope slide, and examine the tissue microscopically. Make observations as to the physical appearance of the callus clumps and the nature of the cells examined microscopically.

Observations:

If you have used callus from soybean cotyledons, continue to subculture at three-week intervals until a sufficient amount can be collected for use in the next exercise.

*Results/
Conclusions*

Present your observations concerning the appearance of the callus clumps and the cells. Did the callus appear to increase in size as a result of cell division or enlargement alone, or as a result of both phenomena? Does the callus contain any differentiated tissue compared to the normal plant part from which it was derived? What are the major physiological differences between so-called normal callus tissue and plant tumor tissue (crown-gall tissue, for example)?

In concluding statements, consider very generally the role of the inorganic salts, sucrose, the vitamins, auxin, and kinetin in the maintenance and growth of callus tissue in culture.

Acknowledgment

The soybean growth medium and general tissue culture techniques used in this exercise were adapted from the work of C. O. Miller.

References

Evans, D. A., W. R. Sharp, P. V. Ammirato, and Y. Yamada (eds.). 1983. *Handbook of Plant Cell Culture,* Vols. 1, 2, 3. New York: Macmillan Publishing Company, pp. 177–227.

Miller, C. O. 1960. An assay for kinetin-like materials. *Plant Physiol.* 35, Supplies XXVI.

Miller, C. O. 1961. A kinetin-like compound in maize. *Proc. Nat. Acad. Sci.* 47: 170–174.

Miller, C. O. 1962. Interaction of 6-Methylaminopurine and adenine in division of soybean cells. *Nature* 194: 787–788.

Miller, C. O. 1963. Kinetin and kinetin-like compounds. In Linskens, H. F. and M. V. Tracy (eds.), *Modern Methods of Plant Analysis.* Berlin: Springer-Verlag, Vol. 6: pp. 194–202.

White, P. R. 1954. *The Cultivation of Animal and Plant Cells.* New York: The Ronald Press Co.

EXERCISE 60

The effect of kinetin on growth of soybean callus tissue in culture: the basis for a bioassay

Introduction

As mentioned previously, the soybean cotyledon callus tissue has been used in the past as a cytokinin bioassay. Since intact cotyledons do not produce cytokinins or auxin, the callus tissue in culture exhibits an absolute requirement for these exogenously applied hormones. These hormones are required for cell division and enlargement and maintenance in culture. For this reason, the soybean cotyledon callus tissue can be used as a very effective cytokinin bioassay. As one might expect from a good bioassay, increasing concentrations of cytokinin incorporated into basic culture medium will stimulate a proportional increase in growth of the tissue over time. The following exercise is designed to show the influence of kinetin on cultured soybean callus growth.

Purpose

To study the effect of different concentrations of kinetin on the growth of callus tissue derived from cotyledons of soybean (*Glycine max*).

Materials

soybean callus tissue (*Glycine max*
L. 'Acme'; other cultivars may
also be used — see Appendix II)
stock solutions (see Table 60–1)
ethanol (95%)
30 g sucrose
10 g agar (powdered or flake)
cotton (nonabsorbent)
Erlenmeyer flasks (one 1 liter; five
500 ml; twenty 125 ml)
volumetric flasks (one 500 ml; five
200 ml; one 100 ml)

graduated cylinder (100 ml)
2 pipets (5 ml, graduated)
5 petri dishes (wrapped in a paper
bag and sterilized)
pH meter
autoclave
transfer hood or room (not absolutely necessary)
Bunsen burner
balance (analytical)
forceps (length: 250 mm)
scalpel

Procedure

A. Basal growth medium

Add approximately 200 ml of distilled water to a 1 l Erlenmeyer flask. Then pipet 5 ml of each of the stock solutions 1 through 11 as indicated in Table 60–1. Do not add kinetin at this time. After mixing thoroughly, add 30 g of sucrose. Bring the volume of the solution to approximately 490 ml with distilled water and adjust the pH to 5.8 with sodium hydroxide. Adjust the final volume to 500 ml with distilled water. This solution will be referred to as *double-strength basal medium*.

B. Test media of various kinetin concentrations

Label five Erlenmeyer flasks (500 ml) and pour 100 ml of the double-strength basal medium into each. As designated in Table 60–2, add the appropriate amount of the kinetin stock solution (solution 12). Check and adjust the pH of each test solution to pH 5.8 if necessary. Then bring the final volume of each to 200 ml with distilled water.

Add 2 g of agar to each flask and subsequently autoclave for five to ten minutes, or appropriately heat to melt the agar. Be sure the agar is completely melted and mixed.

Pour 50 ml of the melted agar medium from each 500 ml Erlenmeyer flask into 125 ml Erlenmeyer flasks correspondingly labeled so that each treatment is represented by four 125 ml flasks.

Stopper the flasks with nonabsorbent cotton plugs and sterilize by autoclaving for fifteen minutes at 121°C and 15 lb./sq. in. pressure. After sterilization, allow the flasks to stand at room temperature until the medium has solidified. Also autoclave five petri dishes wrapped in a paper bag.

TABLE 60–1   Stock solutions for soybean callus tissue cultures*

Stock Solution Number	Component	Amount (g)	Final Aqueous Volume (ml)
1	KH_2PO_4	30.000	500
2	KNO_3	100.000	500
	$Ca(NO_3)_2 \cdot 4H_2O$	50.00	500
3	$MgSO_4 \cdot 7H_2O$	7.150	
	KCl	6.500	500
4	$MnSo_4 \cdot H_2O$	0.492	
	$ZnSO_4 \cdot 7H_2O$	0.380	500
5	H_3BO_3	0.160	
	KI	0.080	500
6	NH_4NO_3	100.000	
7	$EDTA\ Na_2 \cdot 2H_2O$	1.340	
	$FeSO_4 \cdot 7H_2O$	0.990	500
8	$Cu(NO_3)_2 \cdot 3H_2O$	0.035	500
9	$(NH_4)_6Mo_7O_{24} \cdot 4H_2O$	0.010	500
10	Thiamin·HCl	0.080	500
	Nicotinic acid	0.200	
	Pyridoxine·HCl	0.080	500
	Myo-inositol	10.000	
11	α NAA	0.040	100
12	Kinetin	0.020	1000

* The stock solutions as given are generally prepared by investigators performing numerous experiments. If the solutions are to be only for class use, the amount of the components and final solution volumes should be scaled down accordingly.

TABLE 60–2   Soybean growth medium with various kinetin concentrations

Flask Number	Double-Strength Basal Medium (ml)	Distilled Water (ml)	Kinetin Stock (ml)	Final Volume (ml)	Final Kinetin Concentration (mg/l)
1	100	100	0	200	0
2	100	99.5	0.5	200	0.05
3	100	95	5.0	200	0.50
4	100	50	50.0	200	5.00
5	100	0	100.0	200	10.00

C.   Callus planting

If a transfer hood or room is not available, the callus planting procedures may be performed in the laboratory with minimum contamination by adhering to standard aseptic techniques. The arrangement of the necessary equipment and the planting procedure are illustrated (Figure 60–1). Dip the forceps and scalpel used to handle and cut the tissue into 95% ethanol and flame before each operation. With forceps sterilized in this manner, transfer a large piece of stock soybean callus tissue (about 300 mg) to a sterile petri dish and cover immediately. Remove the cover and, using a sterile scalpel, cut the tissue into small equal pieces of approximately 5 mg per piece. Sufficient tissue pieces should be cut in order to plant three pieces per flask; twenty additional pieces will be used as a representative aliquot for initial fresh weight determinations. The mean fresh weight of the aliquot will provide an estimate of the fresh weight per piece originally planted.

After planting, place the cultures on the laboratory bench (since the callus will grow reasonably well at room temperature and under reduced light conditions) or in a growth chamber where they will grow at a temperature of 27°C and be exposed constantly to fluorescent lighting of about 40 ft.-c. for a period of four weeks.

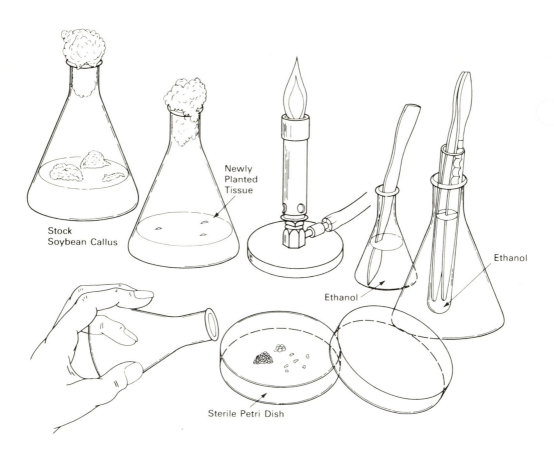

FIGURE 60–1   Equipment for culturing soybean callus.

D.   Callus fresh weight determination and observations

At the end of four weeks' growing time, determine whether the growth increase of the kinetin-treated tissues resulted from cell division, expansion, or both, and determine the fresh weight of the individual pieces.

For fresh weight determinations, carefully remove tissue clumps from the agar and then weigh them on an analytical balance.

Observations and fresh weight determinations:

E.   Statistical treatment of the results

In experiments of this type, it is advantageous to determine the level of probability in which the mean differences between the individual treatments and the controls arise from the activities of exogenously supplied kinetin, as compared to that from inherent variation of the biological material being used. Therefore, perform a simple "t-test" by first filling in the values for all the necessary parameters to be used in Table 60–3.

T-test for equal numbers

(Use if no tissues have been lost due to contamination.) After the parameters for the results of each treatment are calculated, any two means in the experiment may be analyzed statistically by performing a t-test according to either one of the following, whichever is appropriate:

$$ t = \frac{\overline{X}_1 - \overline{X}_2}{(S_{\overline{x}})^2 + (S_{\overline{x}})^2} = \frac{\overline{D}}{(S_{\overline{x}})^2 + (S_{\overline{x}})^2} $$

The difference between the means being compared ($\overline{D}$) is obtained by subtracting the lesser mean weight ($\overline{X}_2$) from the greater mean weight ($\overline{X}_1$). The difference is divided by the square root of the sum of the squared standard error of the means of the treatments being compared.

Complete the necessary calculations for the t-value, then determine the level of significance by finding the corresponding value listed in any standard t-distribution table for the appropriate degrees of freedom (f). For this experiment, the

TABLE 60–3   Statistical determinations

Parameter	Symbol	Determination	Kinetin Concentration (mg/l)				
			0	0.05	0.5	5.0	10.0
Number of individuals/treatment	N	12*	12	12	12	12	12
Degrees of freedom	f	$f = N - 1$	11	11	11	11	11
Sum of the individual weights	ΣX	$\Sigma X = X + X + X + \cdots$					
Total weight squared	$(\Sigma X)^2$	$(\Sigma X)^2 = (\Sigma X)(\Sigma X)$					
Sum of individual weights squared	$\Sigma(X^2)$	$\Sigma(X^2) = X^2 + X^2 + X^2 + \cdots$					
Mean weight/treatment	$\overline{X}$	$\overline{X} = \dfrac{\Sigma X}{N}$					
Correction factor	CF	$CF = \dfrac{(\Sigma X)^2}{N}$					
Sum of individual deviations from the mean squared	Σx^2	$\Sigma x^2 = \Sigma(X^2) - CF$					
Variance	S^2	$S^2 = \dfrac{\Sigma x^2}{N - 1}$					
Standard deviation	S	$S = \sqrt{S^2}$					
Standard error of the mean	$S_{\overline{x}}$	$S_{\overline{x}} = \sqrt{\dfrac{\Sigma x^2}{N(N - 1)}}$					

* N is equal to 12 unless some tissues were lost due to contamination.

total degrees of freedom (N – 1) is equal to 12 – 1 plus 12 – 1, or 22. If the t-value for any two means being compared is in the probability range of 0.05 (5%), the difference between means is said to be significant, and in at least 95% of the cases the growth response is due to the treatment rather than to inherent sample variation. Similarly, if the t-value obtained is in the probability range of 0.01 (1% level), the difference is said to be highly significant.

T-test for unequal numbers

(Use if tissues have been lost by contamination.) During the growth period, several tissues of one treatment may be lost due to contamination. Therefore, the situation might arise in which the mean of the control (– kinetin) is based on N = 12, and the mean of one of the kinetin treatments is based on N = 10. It is advisable to calculate an average standard deviation of the two treaments from the following expression:

$$S = \sqrt{\frac{\Sigma x^2 + \Sigma x^2}{N + N - 2}}$$

The sum of the sum of the deviations from the mean squared (Σx^2) of the two treatments is divided by the sum of the number (N) of individuals in the two treatments being compared minus 2. Find the square root and, using this value (S), calculate the standard error for the mean of each treatment as indicated for the following example:

Treatment 1 (– kinetin): $\dfrac{S}{\sqrt{N_1}} = S_{\bar{x}}$

Treatment 2 (+ kinetin 0.05 mg/l): $\dfrac{S}{\sqrt{N_2}} = S_{\bar{x}}$

With the two adjusted standard errors, calculate the t-value according to the expression previously described:

$$t = \frac{\overline{X}_1 - \overline{X}_2}{(S_{\bar{x}})^2 + (S_{\bar{x}})^2}$$

Using a t-distribution table, determine whether the difference between the two means is statistically significant at the 5% level or highly significant at the 1% level. Remember to use the appropriate degrees of freedom. For example, if a comparison is being performed between a mean based on ten individuals with one based on twelve, the degrees of freedom would be equal to 10 – 1 plus 12 – 1, or 20.

Results/ Conclusions

In the presentation of results, a table may be designed indicating the level of significance between the various kinetin treatments and the control (– kinetin).

Also construct a graph in which the mean fresh weights (mg) are plotted against the respective kinetin concentrations. Discuss the effect of different kinetin concentrations on the growth of soybean callus tissue in culture.

Acknowledgment   This exercise was adapted from the work of C. O. Miller.

References   Blaydes, D. F. 1966. Interaction of kinetin and various inhibitors in the growth of soybean tissue. *Physiol. Plantarum* 19: 748–753.

Miller, C. O. 1960. An assay for kinetin-like materials. *Plant Physiol.* 35, Supplement XXVI.

Miller, C. O. 1961. A kinetin-like compound in maize. *Proc. Nat. Acad. Sci.* 47: 170–174.

Miller, C. O. 1961. Kinetin and related compounds in plant growth. *Ann. Rev. Plant Physiol.* 12: 395–408.

Miller, C. O. 1963. Kinetin and kinetin-like compounds. In Linskens, H. F. and M. V. Tracey (eds.), *Modern Methods of Plant Analysis*, Vol. 6. Berlin: Springer-Verlag, pp. 194–202.

Miura, G. A. and C. O. Miller. 1969. 6-γ,γ-Dimethylallylaminopurine as a precursor to zeatin. *Plant Physiol.* 44: 372–376.

Witham, F. H. 1968. Effect of 2,4-dichlorophenoxylacetic acid on the cytokinin requirement of soybean cotyledon and tobacco stem pith callus tissues. *Plant Physiol.* 43: 1455–1457.

Witham, F. H. and C. O. Miller. 1965. Biological properties of a kinetin-like substance occurring in *Zea mays. Physiol. Plantarum* 18: 1007–1017.

(Also see the references following Exercises 58 and 59.)

EXERCISE 61

The effect of kinetin and indole-3-acetic acid on growth and shoot formation of tobacco callus tissue

Introduction

During early work on the isolation and chemical characterization of kinetin, cultured pith tissue from tobacco stems was used extensively as a bioassay (Miller and Skoog). During the course of this work, it was observed that when the right amounts of kinetin and IAA along with the proper vitamins and minerals and a carbon source (sucrose) are present in the growth medium, the cells of tobacco stem pith tissue divide, enlarge, and produce a loosely arranged mass of undifferentiated cells (callus). The cultured callus from tobacco stem pith, however, requires the presence of a cytokinin and auxin (IAA, NAA, or 2, 4-D) for continued growth and maintenance in culture. Miller and Skoog further demonstrated that when the concentration of kinetin and IAA is changed to a high concentration of kinetin to auxin ratio (higher than required for callus growth), the callus is stimulated to differentiate into shoots and leaves, often giving rise to plantlets. The plantlets may develop further and, with the formation of roots, give rise to intact plants with subsequent growth and formation of seeds. When plantlets are placed back on a medium favoring undifferentiated growth, they will form callus tissue. These observations have led to many of the widespread techniques currently used in the micropropagation of diverse species of plants from cells, tissues, and organs. The techniques demonstrated in the following exercise are similar to those used in earlier work on the establishment and growth of pith tissue isolated from tobacco stems.

Purpose

To establish stem pith callus tissue in culture and to determine the effect of kinetin and auxin upon callus proliferation and shoot formation.

273

Materials

similar material for Parts A and B:

- tobacco plants (*Nicotiana tabacum* L.) about 2 ft. tall
- stock solutions (see Table 61–1)
- ethanol (95%)
- cotton (nonabsorbent)
- graduated cylinders (100 ml)
- 12 pipets (10 ml or 5 ml, graduated)
- 1 pipet (1 ml, graduated)
- 5 petri dishes (wrapped in paper bag and autoclaved)
- pH meter, 1N NaOH, 1N HCl
- autoclave
- transfer hood or room
- Bunsen burner
- forceps
- cork borer
- scalpel
- ruler (cm)
- growth chamber or bench with constant illumination (40 ft.-c.)
- balance

special material for Part A:

- ethanol (70%)
- 30 g sucrose
- 10 g agar (powdered or flake)
- 2 l Erlenmeyer flask
- 20 Erlenmeyer flasks (125 ml)
- 1 l volumetric flask

special material for Part B:

- 60 g sucrose
- 7 g agar (powdered or flake)
- 1 l Erlenmeyer flask
- 7 Erlenmeyer flasks (500 ml)
- 28 Erlenmeyer flasks (125 ml)

Procedure

It is recommended that tobacco callus stock be established in culture and be ready in suitable amounts for the class performance of Part B of the procedure. Part A may be performed in class merely to demonstrate the methods involved in the isolation and establishment of the pith tissue in culture. Another recommendation is that all the media used in the experiment be prepared before the class meeting.

With suitable preparation before class, the entire experiment (Parts A and B) can be set up in one laboratory meeting with emphasis being given to the various techniques relevant to the tissue isolation and planting.

A. Establishment of tobacco stem pith callus tissue in culture

Basal growth medium for tobacco stem pith callus tissue

Add approximately 500 ml of distilled water to a 2 l Erlenmeyer flask. With the exception of solutions 7 (EDTA), 11 (IAA), and 12 (kinetin), pipet 5 ml of the stock solutions (Table 61–1) into the flask. Then add 12.5 ml of solution 7 (EDTA), 50 ml of solution 11 (IAA), and 50 ml of solution 12 (kinetin).

TABLE 61–1   Stock solution for tobacco callus tissue cultures*

Stock Solution Number	Component	Amount (g)	Final Aqueous Volume (ml)
1	KH_2PO_4	30.000	500
2	KNO_3	100.000	500
	$Ca(NO_3)_2 \cdot 4H_2O$	50.000	
3	$MgSO_4 \cdot 7H_2O$	7.150	500
	KCl	6.500	
4	$MnSO_4 \cdot H_2O$	0.492	500
	$ZnSO_4 \cdot 7H_2O$	0.380	
5	H_3BO_3	0.160	500
	KI	0.080	
6	NH_4NO_3	100.000	
7	EDTA $Na_2 \cdot 2H_2O$	1.340	500
	$FeSO_4 \cdot 7H_2O$	0.990	500
8	$Cu(NO_3)_2 \cdot 3H_2O$	0.035	
9	$(NH_4)_6Mo_7O_{24} \cdot 4H_2O$	0.010	500
10	Thiamin · HCl	0.080	
	Nicotinic acid	0.200	
	Pyridoxine · HCl	0.080	500
	Myo-inositol	10.000	
11	IAA	0.020	200
12	Kinetin	0.020	1000

* The stock solutions as given are generally prepared by investigators performing numerous experiments. If the solutions are to be only for class use, the amount of the components and final solution volumes should be scaled down accordingly.

Dissolve 10 g of sucrose in the solution and add sufficient distilled water to bring the volume to approximately 950 ml. Adjust the pH to 5.8 with sodium hydroxide and then the final volume to 1 l with distilled water.

To the above solution, add 10 g agar and subsequently autoclave for five to ten minutes, or heat until the agar is completely melted. Be sure the agar is completely melted and the solution is thoroughly mixed.

Pour 50 ml of the melted agar medium into each of twenty 125 ml Erlenmeyer flasks. Stopper the flasks with nonabsorbent cotton plugs and sterilize by autoclaving for fifteen minutes at 121°C and 15 lb./sq. in. pressure. Wrap five petri dishes in a paper bag and sterilize at the same time as the flasks.

After sterilization, maintain the flasks at room temperature until the medium has solidified. The medium may be used immediately after hardening or stored in a refrigerator for several days.

Preparation and planting of sterile stem pith tissue

If a transfer hood or room is not available, the following procedure may be performed in the laboratory with minimum contamination by adhering to standard aseptic techniques. The arrangement of the necessary equipment and outline of the general technique are shown in Figure 61–1. The forceps, cork borer, and scalpel used to handle and cut the tissue should be dipped into 95% ethanol and flamed before each operation.

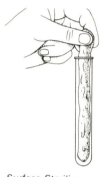

1. Surface Sterilize Tobacco Stem in 95% Ethanol

2. Trim and Discard End of Each Stem Section

3. Punch Out Pith with a Sterile Cork Borer

4. Trim Off and Discard Ends of Pith Tissue and Transfer Sterile Cylinder to Another Sterile Petri Dish

5. Cut Pith Tissue into 1 cm Pieces

6. Cut Each 1 cm Piece into Quarters

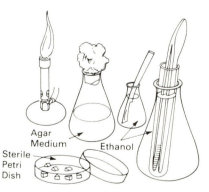

7. Equipment for Planting

8. Pith Tissue Planted

FIGURE 61–1   Tobacco stem pith isolation.

Trim the leaves from the stem of a tobacco plant (approximate height, 2 ft.) and cut several stem sections of about 9 cm in length. Surface-sterilize the sections by dipping in 70% ethanol.

Trim the sections to 7 cm in length by cutting off sections from each end with a sterile scalpel. Then punch out the pith tissue with a sterile cork borer (diameter about the same as the pith tissue). Transfer the pith tissue to a sterile petri dish, trim the ends again, and transfer the sterile cylinder to a fresh petri dish.

With a scalpel, cut each section of pith tissue into smaller sections (about 1 cm wide). Quarter the sections and plant these in each of the flasks containing the basal medium (three pieces per flask). Maintain the cultures in a growth chamber for approximately four to five weeks at 27°C and under constant fluorescent lighting (about 40 ft.-c.). The cultures may be grown in the laboratory under constant illumination if a growth chamber is not available. Subculture the resulting callus tissue until a sufficient amount (about one flask containing three large pieces of callus per student or group) is ready for the subsequent experiment.

B. Effect of kinetin and auxin upon the growth and shoot formation of tobacco callus tissue

Basal growth medium

Add approximately 200 ml of distilled water to a 1 l Erlenmeyer flask. Then pipet 10 ml of each of the stock solutions 1 through 10 as indicated in Table 61–1. Do not add indole-3-acetic acid or kinetin at this time. After mixing thoroughly, dissolve 60 g sucrose in the solution and add sufficient distilled water to bring the volume to approximately 950ml. Adjust the pH to 5.8 with sodium hydroxide and then the final volume to 1 with water. This solution will be referred to as *double-strength basal medium*.

Preparation of test media containing different kinetin concentrations

Label seven Erlenmeyer flasks (500 ml) and pour 100 ml of the double-strength basal medium into each. As designated in Table 61–2, add the appropriate amount of the kinetin and auxin stock solutions. Check and adjust the pH of each solution to pH 5.8 if necessary. Then adjust the volume of the contents of each flask to 200 ml with distilled water.

Add 1 g of agar to each flask and subsequently autoclave for five to ten minutes, or appropriately heat to melt the agar. Be sure the agar is completely melted and mixed.

Pour 50 ml of the melted agar medium from each 500 ml flask into 125 ml Erlenmeyer flasks correspondingly labeled (four flasks per treatment).

Stopper the flasks containing medium with nonabsorbent cotton plugs and sterilize by autoclaving for fifteen minutes at 121°C and 15 lb./sq. in. pressure. Also autoclave five petri dishes wrapped in a paper bag. After sterilization, maintain the flasks at room temperature until the medium has solidified.

TABLE 61–2   Tobacco callus growth medium with various kinetin and auxin concentrations

Flask Number	Double-Strength Basal Medium (ml)	Kinetin Stock (ml)	IAA Stock (ml)	Distilled Water (ml)	Final Volume (ml)	Kinetin Concentration (mg/l)	IAA Concentration (mg/l)
1	100	0	0	100	200	0	0
2	100	0	10	90	200	0	5.0
3	100	5.0	0	95	200	0.5	0
4	100	0.5	10	89.5	200	0.05	5.0
5	100	5.0	10	85	200	0.5	5.0
6	100	10.0	10	80	200	1.0	5.0
7	100	50.0	10	40	200	5.0	5.0

Callus planting

Be sure to use aseptic technique and sterilize the forceps and scalpel used to handle and cut the tissue before each operation. With forceps, transfer a large piece of tobacco callus stock to a sterile petri dish. Then cut the tissue into small pieces (about 40 mg fresh weight). Plant these pieces in each of the 125 ml Erlenmeyer flasks containing the test medium.

After planting, place the cultures on the bench under constant lighting or in a growth chamber where they can be maintained at 27°C and constant fluorescent lighting (about 40 ft.-c.) for a period of five to eight weeks.

During the growth period, examine the cultures (do not remove the plugs) and make weekly notes as to the appearance of the tissues.

Observations:

After the growth period, record your final observations concerning the general appearance of the tissues grown on the different test media. Note the color of the tissues, the relative amounts of callus tissue present, and the average number of shoots per piece. Determine the average fresh weight of the tissues in each treatment. Also, examine a thin slice of the callus tissue microscopically to determine whether growth was due to cell division or enlargement or both. Design a suitable table and record your final observations and measurements. As part of the table, you should indicate the treatment number, the relative amount of callus growth, the average number of shoots formed, and the average fresh weight of the pieces in each treatment.

Results/
Conclusions

Interpret the results (as presented in your table) and base your conclusions in part on answers to the following:

1. In the flasks where there was appreciable callus proliferation, did the increase in size result entirely from cell division or enlargement?

2. Did kinetin or auxin alone stimulate growth or differentiation of the tobacco callus?

3. Is it possible to determine from this experiment which hormone primarily stimulated cell division? Explain.

4.   What experimental advantages does the use of tobacco callus cultures have in studies pertaining to the chemical regulation of growth?

5.   Is the tobacco callus tissue culture system useful as a means of bioassay for naturally occurring cytokinins or auxins? Explain.

Acknowledgment   This exercise was adapted for class use from the work of F. Skoog and colleagues.

References

Horgan, R. 1984. Cytokinins. In M. B. Wilkins (ed.). *Advanced Plant Physiology.* London: Pitman Publishing, Limited, pp. 53–75.

Letham, D. S. 1978. Cytokinins. In Letham, D. S., P. B. Goodwin, and T. J. V. Higgins (eds.), *Phytohormones and Related Compounds: A Comprehensive Treatise.* Amsterdam: Elsevier/North-Holland Biomedical Press, pp. 205–263.

Linsmaier, E. M. and F. Skoog. 1965. Organic growth factor requirements of tobacco tissue cultures. *Physiol. Plantarum* 18: 100–127.

Miller, C. O., F. Skoog, F. S. Okumura, M. H. Von Saltza, and F. M. Strong. 1955. Structure and synthesis of kinetin. *J. Amer. Chem. Soc.* 77: 2662–2663.

Miller, C. O., F. Skoog, F. S. Okumura, M. H. Von Saltza, and F. M. Strong. 1956. Isolation, structure, and synthesis of kinetin, a substance promoting cell division. *J. Amer. Chem. Soc.* 78: 1375–1380.

Miller, C. O., F. Skoog, M. H. Von Saltza, and F. M. Strong. 1955. Kinetin, a cell division factor from deoxyribonucleic acid. *J. Amer. Chem. Soc.* 77: 1392.

Murashige, T. and F. Skoog. 1962. A revised medium for rapid growth and bioassay with tobacco tissue cultures. *Physiol. Plantarum* 15: 473–497.

Rogozinska, J. H., J. P. Helgeson, and F. Skoog. 1964. Tests for kinetin-like growth promoting activities of triacanthine and its isomer, 6-(γ,γ-dimethylallylamino) purine. *Physiol. Plantarum* 17: 165–178.

Skoog, F. and C. O. Miller. 1957. Chemical regulation of growth and organ formation in plant tissues cultured *in vitro. Symp. Soc. Exp. Biol.* 11: 118–131.

The effect of kinetin on chlorophyll retention in detached wheat leaves

Introduction

Plant scientists have observed for some time that when a healthy green leaf is removed from a plant there is a rapid loss of chlorophyll. We now know that beginning with the detachment of the leaf there is a breakdown of proteins in the blade, which is followed by a flow of non-protein nitrogen, lipid, and nucleic acid components into the petioles.

Chibnall first demonstrated in the early 1950s that if roots were formed on the petiole of detached leaves (induced by IAA treatment), the above-mentioned symptoms of senescence were retarded or completely inhibited. He suggested from these observations that the roots on detached leaves produce a substance, a hormone, that is translocated to the lamina and acts to retard senescence. Later, Richmond and Lang showed that when detached leaves are treated with kinetin protein, degradation and chlorophyll loss are retarded.

As mentioned in the introduction to Exercise 56, gibberellins are also involved in delaying the protein and chlorophyll changes of detached leaves.

Purpose

To study the influence of cytokinins in delaying senescence, as evidenced by chlorophyll retention in detached wheat leaves.

Materials

60 nine-day-old wheat seedlings
(*Triticum aestivum* L. 'Red
Coat')
acetone (85 %)
8 beakers (1000 ml)
8 petri dishes

3 separate kinetin solutions
(10 mg/l, 5 mg/l, and 0.5
mg/l)
Waring blender or mortar and
pestle
colorimeter or spectrophotometer

Procedure

A.   Plant material

Sow wheat seeds (*Triticum aestivum* L. 'Red Coat') in a flat of soil and maintain in the greenhouse until the plants are about 10 cm tall (about nine days old). Then detach fifty leaves at the ligule and place them together in a beaker of distilled water.

Maintain the leaves in distilled water and under fluorescent light for two to three days. This aging period will allow for the initiation of those processes leading to observable chlorophyll degradation.

If wheat is not available, whole leaves or leaf disks from tobacco, sunflower, cocklebur, or other thin-leave species may be used. Also, if mature aging leaves (pale green in color) are used, the presoaking described for the wheat leaves may be omitted. However, if leaf disks are used, line the petri dishes with a pad of filter paper and moisten with about 1 ml of test solution (see Part B). If excess liquid is used, the disks tend to blacken with time due to waterlogging of the tissues. Also, along with the disks, wrap the dishes in moistened paper towels or store in a humid environment between observations.

B.   Effect of kinetin on chlorophyll retention in detached leaves

Prepare eight petri dishes (two per treatment) containing 25 ml of each of the following:

Treatment 1: distilled water
Treatment 2: kinetin solution (0.5 mg/l)
Treatment 3: kinetin solution (5 mg/l)
Treatment 4: kinetin solution (10 mg/l)

From the beaker of distilled water, transfer five detached leaves selected at random to each petri dish. Cover the dishes and place them under fluorescent lights. Now determine the total chlorophyll content (mg chlorophyll/g of tissue) of the leaves remaining in the beaker (see Exercise 29). The total chlorophyll value obtained initially will provide a reference for comparison with the experimental leaves later.

During the course of the experiment (eight to ten days), expose the detached leaves in the petri dishes to sixteen hours of darkness and a temperature of 4 to 10°C, and to eight hours of light at a temperature of 20 to 25°C each day. Other methods may be used to inhibit microbial growth, but the sixteen-hour cold treatment each day is adequate and eliminates the need for surface-sterilization of the leaves and of sterile solutions and petri dishes. However, critical work would require greater precautions than outlined here.

At daily intervals, observe the color of each detached experimental leaf and rate it from 1 to 5 in comparison to the color of normal intact leaves rated at five.

After eight to ten days, determine the total chlorophyll content (mg chlorophyll/g tissue) of the leaves in each treatment.

Observations and total chlorophyll values:

Results/
Conclusions

Construct a table indicating your ratings based on the color of the leaves in each treatment and the number of days after the start of the experiment. Also construct a graph with the total chlorophyll values plotted against the four treatments. If desired, a bar graph may be used to accomplish the illustration of results.

Interpret the results on the basis of the effect of kinetin upon the retention of chlorophyll by detached wheat leaves. In concluding remarks, indicate some of the other general biological responses attributed to the action of cytokinins. With respect to your results, present a possible biochemical explanation for the effect of cytokinins on detached leaves.

Acknowledgment

This exercise was adapted from the work of C. Person, D. J. Samborski, and F. R. Forsyth.

References

Chibnall, A. C. 1954. Protein metabolism in rooted runner-bean leaves. *New Phytologist* 53: 31–35.

Letham, D. S. and L. M. S. Palni. 1983. The biosynthesis and metabolism of cytokinins. *Ann. Rev. Plant Physiol.* 34: 163–197.

Osborne, D. J. 1962. Effect of kinetin on protein and nucleic acid metabolism in *Xanthium* leaves during senescence. *Plant Physiol.* 37: 595–602.

Person, C., D. J. Samborski, and F. R. Forsyth. 1957. Effect of benzimidazole on detached wheat leaves. *Nature* 180: 1294–1295.

Richmond, A. E. and A. Lang. 1957. Effect of kinetin on protein content and survival of detached *Xanthium* leaves. *Science* 125: 650–651.

Sugiura, M., K. Umemura, and Y. Oota. 1962. The effect of kinetin on protein levels of tobacco leaf disks. *Physiol. Plantarum* 15: 457–464.

Thomas, H. and J. L. Stoddart. 1980. Leaf senescence. *Ann. Rev. Plant Physiol.* 31: 83–111.

Witham, F. W. and C. O. Miller. 1965. Biological properties of a kinetin-like substance occurring in *Zea mays*. *Physiol. Plantarum* 18: 1007–1017.

EXERCISE 63

The interaction of kinetin and auxin on secondary shoot development

Introduction

Apical dominance, a widely observed phenomenon in many plants, is seemingly due to the action of auxin on the lateral buds. It is a curious anomaly that the apical bud, the site of auxin synthesis, will grow while lateral buds are inhibited by auxin translocated from the stem tip. The explanation, in part, may be due to the accumulation of relatively high concentrations of auxin in the lateral buds compared with that in the apical bud. Further, it is likely that auxin promotes ethylene synthesis in lateral buds, which in turn inhibits bud growth and development.

When the apical bud is damaged by pathogens or removed, the lateral bud closest to the apex grows and, assuming the apical position, exerts apical dominance over the remaining lateral buds. If, however, a renewed apical bud area or the lateral buds of a de-tipped plant are treated with auxin, apical dominance is maintained on all buds.

Kinetin applications will stimulate lateral bud development and by some unknown action will overcome apical dominance (see Wickson and Thimann 1958). Other cytokinins produce the same response of promoting lateral bud growth (see Witham and Miller 1965).

Samuels (1961), working with intact seedlings of peas (*Pisum sativum*, 'Alaska') also demonstrated that lateral bud inhibition was overcome to varying degrees by the application of different concentrations of applied kinetin.

This exercise is designed to illustrate the influence of indole-3-acetic acid and kinetin on lateral bud growth and development.

Purpose

To demonstrate that the lateral bud inhibition, natural or induced by exogenous indole-3-acetic acid, can be overcome to varying degrees by different concentrations of applied kinetin.

284

Materials

50 pea seeds (*Pisum sativum* L. 'Alaska')

100 ml Clorox solution (10%)

3 separate kinetin solutions (20 mg/l; 10 mg/l; 1 mg/l)

indoleacetic acid solution (20 mg/l)

3 separate stock solutions containing:
— kinetin (10 mg/l) + IAA (20 mg/l)
— kinetin (10 mg/l) + IAA (10 mg/l)
— kinetin (10 mg/l) + IAA (1.0 mg/l)

autoclaved distilled water

3 deep petri dishes or beakers, covered, and containing sterile distilled water

sterile petri dishes containing three pads of Whatman No. 1 filter paper wetted with 10 ml of sterile distilled water

pipets (10 ml)

8 test tubes (16 cm × 15 mm)

test tube racks

forceps (sterilized before each operation)

Procedure

A. Seed germination under aseptic conditions

Soak Alaska pea seeds in 10% Clorox for five to ten minutes. Aseptically transfer the seeds from the Clorox solution with sterile forceps to a beaker containing sterile distilled water. After five to ten minutes, transfer the seeds to a second beaker containing sterile distilled water, and then to a third. After the three separate water washes, transfer the seeds aseptically to sterile petri dishes containing three pads of Whatman No. 1 filter paper wetted with approximately 10 ml of water. Cover the dishes and allow the seeds to germinate at room temperature under constant illumination of approximately 40 ft.-c. The seeds are germinated in the light to retard internode elongation of the developing seedlings.

B. Preparation of the treatments

Pipet 10 ml of water or stock solution into 16 cm × 15 mm test tubes so that the test solutions indicated in Table 63–1 are represented. For an individual experiment, prepare five tubes per treatment unless a sufficient number of students are performing the experiment for subsequent pooling of all the results.

Autoclave the tubes and the remainder of the test solutions in separate flasks stoppered with cotton for fifteen minutes at 121°C and 15 lb./sq. in. pressure. After sterilization, allow the solutions to cool to room temperature.

C. Seedling transfer to the various test solutions

Aseptically transfer a three-day-old seedling to each 16 cm × 15 mm test tube containing the sterilized test solutions. Use sterile forceps (dip in alcohol and flame before each operation) to accomplish the transfer. The depth of the solu-

TABLE 63–1   Effect of kinetin and IAA on lateral bud growth

Tube Number	Treatment (mg/l)	Total Number (a) and Mean Number (b)					
		Lateral Shoots		Basal Thickening 1° Stem		Basal Thickening 2° Stem	
		a	b	a	b	a	b
1	Distilled water						
2	Kinetin (1.0)						
3	Kinetin (10.0)						
4	Kinetin (20.0)						
5	IAA (20.0)						
6	Kinetin (10.0) + IAA (1.0)						
7	Kinetin (10.0) + IAA (10.0)						
8	Kinetin (10.0) + IAA (20.0)						

tions in each tube should be sufficient to cover the first nodal region of the seedlings (see Figure 63–1). Allow the seedlings to grow under the same light and temperature conditions as for germination for a period of eight days.

During the course of the experiment, further additions of autoclaved water or the hormone solutions should be made to each tube approximately every forty-eight hours to maintain the first nodal region of each seedling continually immersed.

Observe the seedlings daily for eight days and note the number of slightly developed and fully developed lateral buds and the basal thickening of primary and lateral shoots. Note any gross abnormalities in seedling morphology.

Observations:

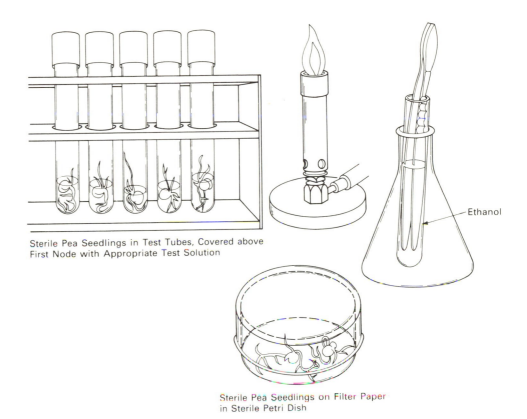

Sterile Pea Seedlings in Test Tubes, Covered above
First Node with Appropriate Test Solution

Ethanol

Sterile Pea Seedlings on Filter Paper
in Sterile Petri Dish

FIGURE 63–1   Transfer of pea seedlings to test tubes.

**Results/
Conclusions**

At the end of the eight days, record the total and mean number of lateral shoots
(regardless of length) for the seedlings of each treatment in Table 63–1. The table
may be used for total class results if indicated by the instructor.

A bar graph based on the mean number of lateral shoots found in each
treatment may also be constructed and included with your observations of seedling morphology in the results section of the report.

From the results obtained, discuss the relationships of kinetin and auxin in
the regulation of secondary shoot development and normal growth. Consider the
phenomenon of auxin-imposed apical dominance and present your own reasons
for the action of kinetin in stimulating lateral bud growth.

Acknowledgment

This exercise was adapted from the work of R. M. Samuels.

References

Letham, D. S. and L. M. S. Plani. 1983. The biosynthesis and metabolism of cytokinins. *Ann. Rev. Plant Physiol.* 34: 163–197.
Miller, C. O. and F. Skoog. 1953. Chemical control of bud formation in tobacco
stem segments. *Amer. J. Bot.* 40: 768–773.

Sachs, T. and K. V. Thimann. 1964. Release of lateral buds from apical dominance. *Nature* 201: 939–940.

Samuels, R. M. 1961. Bacterial induced fasciation in *Pisum sativum* var. *Alaska*. Ph.D. thesis, Indiana University.

Skoog, F. and K. V. Thimann. 1934. Further experiments on the inhibition of the development of lateral buds by growth hormone. *Proc. Natl. Acad. Sci.* 20: 480–485.

Wickson, M. and K. V. Thimann. 1958. The antagonism of auxin and kinetin in apical dominance. *Physiol. Plantarum* 11: 62–74.

Wickson, M. and K. V. Thimann. 1960. The antagonism of auxin and kinetin in apical dominance. II. The transport of IAA in pea stems in relation to apical dominance. *Physiol. Plantarum* 13: 539–554.

Witham, F. H. and C. O. Miller. 1965. Biological properties of a kinetin-like substance occurring in *Zea mays. Physiol. Plantarum* 18: 1007–1017.

The effect of kinetin and gibberellic acid on radish cotyledon enlargement: the basis for a bioassay

Introduction

Cellular enlargement, a biological response usually associated with IAA and gibberellins, is also induced by cytokinins. Of special interest is the cytokinin- and gibberellin-stimulated enlargement of excised radish cotyledons. In addition, kinetin stimulates enlargement of excised cotyledons of numerous plants, including pumpkin, cocklebur, flax, and fenugreek.

In radish cotyledons, the cytokinin-stimulated enlargement is due to cellular enlargement and not cell division. Further, the enlargement is due primarily to stimulated water uptake, which is in response to the production of simple sugars in the cotyledonary cells. In the presence of cytokinin, the sugars appear to build up as lipids decrease. Kinetin does not appear to affect the lipid conversion enzymes, but does stimulate invertase activity (Howard and Witham 1983). In this regard, sucrose, which initially builds up in the cotyledons from the lipid conversions, may be rapidly hydrolyzed to the osmotically active sugars glucose and fructose causing water uptake.

In cytokinin-treated radish cotyledons, zeatin seems to promote cell-wall changes in which cell-wall modification (increased plasticity) takes place. Hence, cytokinins stimulate radish cotyledon enlargement by their action on at least two physiological processes.

The stimulation of excised radish cotyledons by gibberellic acid has not been studied extensively, though the use of the system as a cytokinin bioassay of crude extracts could be complicated by the presence of gibberellins. Nevertheless, the system illustrated in the following exercise is an excellent one for studying the action of either the cytokinins or gibberellins.

Purpose

To investigate the effect of kinetin and gibberellic acid upon radish cotyledon enlargement, and the conditions under which the biological response may be used to assay for synthetic and naturally occurring cytokinins.

Materials

200 radish seedlings (*Raphanus sativus* L. 'Long Scarlet'; market strain is recommended; see Procedure, Part A)

500 ml potassium phosphate buffer (2mM; pH 5.9; see Appendix II)

10 ml kinetin stock solution (5 mg/l water)

10 ml kinetin stock solution (5 mg/l potassium phosphate buffer; pH 5.9)

10 ml gibberellic acid stock solution (5 mg/l water)

10 ml gibberellic acid stock solution (5 mg/l potassium phosphate buffer; pH 5.9)

18 petri dishes lined with a pad of Whatman No. 1 filter paper (9 cm diameter)

Procedure

A. Radish seed germination

Sow 200 radish seeds in a tray lined with filter paper well-wetted with distilled water. Enclose the tray in a plastic bag and maintain the seed in darkness at 26°C. After about thirty hours, excise the smaller cotyledon from each seedling. Be sure to remove all hypocotyl from the excised cotyledons. For the following procedure, select cotyledons of uniform size.

B. Incubation of radish cotyledons in aqueous solutions of kinetin and gibberellic acid

Label each of nine petri dishes containing a pad of Whatman No. 1 filter paper (diameter of 9 cm). Wet each filter paper pad with 3 ml of the appropriate test solution as indicated in Table 64–1.

TABLE 64–1 Effect of kinetin and gibberellic acid on radish cotyledon enlargement (Parts B and C)*

Petri Dish Number	Compound Used*	Part B Concentration in Water (mg/l)	Part C Concentration in Phosphate Buffer (mg/l)	Cotyledon Weight (mg) B	Cotyledon Weight (mg) C
1	Distilled water	—	—		
2	Kinetin	0.005	0.005		
3	Kinetin	0.05	0.05		
4	Kinetin	0.5	0.5		
5	Kinetin	5.0	5.0		
6	Gibberellic acid	0.005	0.005		
7	Gibberellic acid	0.05	0.05		
8	Gibberellic acid	0.5	0.5		
9	Gibberellic acid	5.0	5.0		

* Use the stock kinetin and gibberellic acid solutions to make serial dilutions for the above concentrations of kinetin and gibberellic acid.

Place ten cotyledons on the filter paper in each dish. Cover the dishes and place them in a tray lined with water-moistened paper toweling. Then place the tray in a plastic bag to maintain high humidity conditions. After incubating for three days at 25°C and under constant fluorescent lighting, weigh the cotyledons to the nearest milligram. Record the individual weights for each treatment and enter the average weight for each treatment in Table 64–1.

Cotyledon weights after incubation:

C.   Cotyledon weights after incubation in kinetin or gibberellic acid dissolved in 2mM potassium phosphate buffer (pH 5.8 to 6.0)

Perform exactly the same procedures as outlined in Part B, but this time use the potassium phosphate buffer alone (control) and kinetin or gibberellic acid dissolved in buffer (see Table 64–1 for concentrations). After incubation for three days at 25°C and under constant fluorescent lighting, record the individual weights of the cotyledons in each treatment and enter the average weight for each treatment in Table 64–1.

Cotyledon weights after incubation:

Results/
Conclusions

Present a graph in which the average weights (in mg) of the cotyledons in each treatment are plotted on the ordinate against the various concentrations of kinetin and gibberellic acid (in water and in 2mM phosphate buffer).

As part of your interpretation of the results, indicate the specificity of the test to kinetin dissolved in water and in phosphate buffer. Under what circumstances would you use a relatively pure cytokinin–aqueous preparation in the radish cotyledon test? Under what circumstances would you use a phosphate buffered extract containing the cytokinins?

Acknowledgment This exercise was adapted from the work of D. S. Letham.

References

Banerji, D. and M. Laloraya. Expansion of isolated pumpkin cotyledons with kinetin. *Naturwissenschaften* 52: 349.

Bewli, I. S. and F. H. Witham. 1976. Characterization of the kinetin-induced water uptake by detached radish cotyledons. *Bot. Gaz.* 137: 58–64.

Burrows, W. J. 1975. Mechanism of action of cytokinins. *Current Adv. Plant Sci.* 7: 837–847.

Howard, H. F. and F. H. Witham. 1983. Invertase activity and the kinetin-stimulated enlargement of detached radish cotyledons. *Plant Physiol.* 73: 304–308.

Huff, A. K. and C. W. Ross. 1975. Promotion of radish cotyledon enlargement and reducing sugar content by zeatin and red light. *Plant Physiol.* 56: 429–433.

Letham, D. S. and L. M. S. Palni. 1983. The biosynthesis and metabolism of cytokinins. *Ann. Rev. Plant Physiol.* 34: 163–197.

(Also see the references following Exercises 60, 61, and 62.)

APPENDIX I

Aids for scheduling exercises

The exercises in this manual differ considerably in content and duration and cannot all be completed in a one-semester course. In addition, the differences in class size, available facilities, and course content dictate flexibility rather than the strict limitations imposed by a prescribed schedule of performance. Therefore, specific aids for scheduling are presented here so that individuals may coordinate selected exercises according to their own requirements.

Generally, the time necessary for an experimental setup, and in some cases for the completion of an entire exercise, will not exceed two two-hour class meetings. However, there are numerous exercises that require observations and experimental manipulation to be completed outside the regularly scheduled class meeting. Initial plans concerning the extent of preparation before class, the assignment of students in groups, and the time involved in the growth and incubation of experimental plant material will significantly affect the successful performance and integration of selected exercises. Therefore, the following aids for scheduling should be used as the basis for coordinating exercises so as to provide a meaningful laboratory experience.

The aids for scheduling have worked well in the past but may be altered according to individual requirements and available facilities. For example, the indicated number of students assigned to work as a group appears to be optimal for a laboratory section of twenty-four members but should be changed accordingly for smaller sections. In addition, the aids should not be used exclusively but only as an index of some situations to be aware of in performing and coordinating exercises.

All the materials indicated for each exercise, including the plants, should be ready before the formal class meeting. The instructions for growing plants and preparing stock solutions, and so on are included in each exercise for purposes of continuity. The student, however, is not expected to use valuable class time for preparations that merely support the pedagogically valuable aspects of an exercise. Another reason for including preparation instructions within each exercise is based on the observation that most students gain deeper insight into the principles of an experiment if they are aware of the supporting activities that they were not necessarily required to perform. On the other hand, for some preparations before class it may be worthwhile, although not absolutely necessary, to expose the student to a particular technique. This aspect of planning is left entirely to the discretion of the instructor.

The duration of various activities outside class emphasizes the fact that for some exercises the student is expected to make minor manipulations of experimental material and take observations outside the regularly scheduled class time. In this respect, it is important to note that the exercises are presented according to content and that this organization does not represent the order in which the exercises should be performed. Some exercises near the end of the manual, particularly those dealing with tissue-culture techniques and photoperiodism, require observations to be taken over several weeks following the initiation of the exercise and should be started rather early in the course. Therefore, considerable attention should be given to the following aids and the details of each exercise for coordinating the laboratory with lecture material.

APPENDIX TABLE 1   Aids for scheduling exercises

Exercise Number	Number of Students in Groups	Preparation of Materials and/or Procedural Part before Class*	Duration of Various Activities outside Class		
			Activity	Procedural Part	Time**
1	2 to 3	—	—	—	—
2	3	Tubes of solidified agar and methyl red indicator (see Part B)	Incubation of liquids	C	1 wk.
3	2 to 4	—	Incubation of tissue	C	1 hr.
4	2	—	—	—	—
5	2	—	Incubation of tissue	Procedure	6 to 24 hr.
6	2	—	Incubation of tissue	B	2 hr.
7	2	Assembling the apparatus (see Part A***)	Assembling the apparatus	A	3 to 4 hr.
8	2	Potted plants	Root pressure measurements	A	2 days
9	2	Pressure bomb apparatus and mature plants for leafy shoots (see Part B)	Transpirational pull measurements	B	2 days
			—	—	—
10	2 to 4	Potted plants	—	—	—
11	2	Potted plants	Measurements	Procedure	8 hr. to several days
12	2	Cobalt chloride paper (see Part A) Potted plants in darkness for 8 hr. (see Part D)	—	—	—
13	2	Potted plants	—	—	—
14	2	Potted plants	—	—	—
15	2	—	Duration and observations	A	2 hr.
			Apparatus maintained	B	24 hr.
16	2	Plants at suitable stages of maturation	—	—	—

Continued

APPENDIX TABLE 1   Continued

Exercise Number	Number of Students in Groups	Preparation of Materials and/or Procedural Part before Class*	Activity	Procedural Part	Time**
17	4	Plants, 7 to 10 days old (see Parts A, B, C)	Growing plants	F	5 wk.
			Dry weight determinations (optional)	F	24 hr.
18	2	—	Dry weight determinations	A	24 hr.
			Ashing dried material	A	2 hr.
19	3	—	Incubation of mixtures	G	2 hr.
20	2	—		—	—
21	2 to 4	Part A	Chromatogram development	B	9 to 12 hr.
22	4	Germinating seeds for 24 hr. (see Parts C, D, E)	Incubation	D	24 hr.
23	3	Part A (chromatographic chambers)	Chromatogram development	A	15 hr.
24	4	—	Chromatogram development and observations	B	1 to 2 hr.
25	3	Germinating corn grains (see Part A)		—	—
		Test reagents (see Materials)			
26	4	Part B	Chromatogram development (phenol)	C	10 to 12 hr.
			Chromatogram development (butanol, acetic acid, water)	C	10 to 12 hr.
27	4	—		—	—
28	2	—	Chromatogram development	B	1 to 2 hr.
29	2 to 3	—		—	—
30	4	—		—	—
31	4	Plants at suitable stages of maturation (see Part A)	Paper chromatogram development (butanol, acetic acid, water)	B	12 hr.
		Thin-layer plates (see Part C***)	Paper chromatogram development (acetic acid)	B	1 to 2 hr.

No.	Students	Material / Preparation	Operation	Procedure	Time
32	2	Nasturtium seeds soaked for several hours in water (see Parts B and C)	Drying thin-layer plates*** or Activating thin-layer plates***	C / C	24 hr. / 35 min.
		Germinating seeds for 48 hr. (see Part C)	Thin-layer chromatogram development	C	1 to 2 hr.
		—	—	—	—
33	2	—	Incubation of mixtures	A	24 hr.
			Enzyme extraction	C-2	1½ hr.
34	2 to 3	Part A***	Incubation of mixtures	B	1 hr.
35	2 to 3	Part A	—	—	—
36	4	Avena seedlings, 4 days old (see Part A)	Incubation of mixtures	D	1 hr.
		Enzyme extraction and preparation (see Parts B and C***)	Incubation of mixtures	E	1 hr.
			Incubation of mixtures	F	1 hr.
			Incubation of mixtures	G	1 hr.
			Incubation of mixtures	H	1 hr.
37	2 to 3	—	—	—	—
38	3	Leaves soaked in tap water for 1 hr. (see Part A)	—	—	—
39	2 to 3	—	—	—	—
40	2	Enzyme preparation (see Part A***)	—	—	—
41	2 to 3	Small potted plants, C_3 and C_4 metabolism (see Part A)	—	—	—
		Open infrared gas exchange system			
42	2 to 4	Seed treatment for 3 days at 37°C	Seed germination	Procedure	17 to 24 hr.
43	2 to 4	Heat treatment of lettuce seed at 35 to 37°C for 3 days (see Part A)	Seed imbibing time	B	16 hr.
			Seed incubation after imbibition	B	32 hr.
44	4	Cocklebur plants, 60 days old (see Part A)	Photoinduction	B	5 days
			Photoinduction	C	5 days

Continued

APPENDIX TABLE 1   Continued

Exercise Number	Number of Students in Groups	Preparation of Materials and/or Procedural Part before Class*	Activity	Duration of Various Activities outside Class	
				Procedural Part	Time**
45	2 to 4	Avena seedlings, 3 days old (see Part A)	Photoinduction	D	5 days
			Results after start of experiment	Results 1	10 days
			Results after start of experiment	Results 2	4 to 5 wk.
			Incubation of segments	C	18 to 20 hr.
46	2	Pea seedlings, 3 to 4 days old (see Parts A and B)	Incubation of seedlings	B	2 to 3 days
		Pea seedlings, 7 to 10 days old (see Parts A and C)	Stem growth of potted plants	C	1 wk.
		Bean seedlings, 7 to 10 days old (see Parts A and D)	Root development of bean plants	D	1 wk.
		Cucumber and rye seed	Seed incubation	E	5 to 7 days
47	2	Seeds soaked for 1 hr. (see Part A)	Unilateral light treatment	B	48 hr.
		Seedlings, 2 to 3 days old (see Part A)	IAA treatment	C	48 hr.
48	2	Germinating corn grains (see Part A)	Gravitropic response	B	1 wk.
			Gravitropic response	C	1 wk.
49	2 to 4	Germinating seed for 72 hr. (see Parts A and B)	Soaking of sections	B	1 hr.
			Incubation of sections	B	20 hr.
50	4	Pea seedlings, 8 to 10 days old (see Part A)	Incubation of sections	C	6 to 12 hr.
51	4	Mung bean seedlings, 8 to 9 days old (see Part A)	Incubation of cuttings	B	5 days
52	2 to 4	Oat seedlings, 5 days old (see Materials)	—	—	—
53	2	Corn seedlings, 3 days old (see Part A)	—	—	—
54	4	Pea seedlings, 7 to 9 days old (see Part A)	Reaction time	C	50 min.

No.	Qty	Material	Activity	Code	Time
		Part B***	Salkowski test	D	30 min.
55	2	Dwarf pea seedlings, 2 wk. old (see Part A-1)	Growth of treated pea plants	B	3 to 4 wk.
		Potted bean plants, 2 to 3 wk. old (see Part A-2)	Growth of treated bean plants	C	3 to 4 wk.
56	4	Barley grains soaked overnight (see Part A)	Incubation of GA_3-treated seeds	C	24 hr.
57	2	Potted *Taraxacum* plants (see Part A)	Leak disk incubation	C	4 days
58	2	Carnations, vegetative shoots 6 to 7 in. long (see Part A)	Incubation of cultivars	B	3 to 4 wk.
59	4	Part A	Seed germination in flasks	C-1	2 wk.
		Part B	Tissue grown in culture	D	3 wk.
		Part C-1 or C-2***	Subculture callus**	D	Every 4 wk.
			or		
			Culture tissue from intact plant	C-2	3 wk.
			Callus grown in culture	D	3 wk.
			Subculture callus***	D	4 wk.
60	2 to 4	Parts A and B	Growth of tissue cultures	C	4 wk.
61	2 to 4	Basal growth medium (see Part A) Stem pith isolation (see Part A***) Part B	Growth of callus	B	5 to 8 wk.
62	2	Wheat seedlings, 9 days old (see Part A)	Incubation of detached leaves	B	8 to 10 days
63	2 to 4	Pea seedlings, 30 hr. old (see Part A)	Pea seedling incubation	C	8 days
64	2	Radish seedlings, 30 hr. old (see Part A)	Cotyledon incubation	B	3 days
			Cotyledon incubation	C	3 days

* Indicates only the time-consuming preparation of plant material and procedural parts of an experiment. Although not indicated in the table, all the materials outlined at the beginning of each exercise should be ready before class.
** The length of the laboratory period and individual planning will determine whether some of these activities can be completed during class time.
*** May be included in the preparations or performed by the student during class.

APPENDIX II

Preparation of materials and reagents

The recipes for the preparation of chemical reagents and explanations of various supplies required for some exercises are listed alphabetically. The exercise number for which the material is required is also provided.

Aluminum chloride spray reagent for Exercise 31

Dissolve 2 g aluminum chloride ($AlCl_3$) and dilute to 100 ml with distilled water.

Ammonium molybdate reagent (freshly prepared) for Exercise 18

Dissolve 7 g ammonium molybdate in 50 ml distilled water. Warm slightly to hasten solution. Filter the solution and pour 50 ml concentrated HNO_3 (sp. g. 1.42) into the filtrate. Then slowly pour the resulting solution into 100 ml distilled water. Store in a drop bottle with a glass stopper. This solution is corrosive, and contact with skin and clothing should be avoided.

Anthrone reagent (freshly prepared) for Exercise 20

Place 0.4 g anthrone in 100 ml concentrated sulfuric acid (H_2SO_4). Mix thoroughly. Avoid contact with skin and clothing.

Barfoed's reagent for Exercise 19

Dissolve 13.3 g crystalline copper acetate in 200 ml distilled water. Filter if necessary and then add to the solution 1.9 ml glacial acetic acid. Mix and use as directed.

Benedict's reagent for Exercises 15, 19, 22

Prepare two separate stock solutions:

    1.   Dissolve 173 g sodium citrate and 100 g anhydrous sodium carbonate (Na_2CO_3) in 800 ml warm distilled water. Filter and dilute the filtrate to 850 ml with distilled water.

2.   Dissolve 17.3 g pure crystalline copper sulfate in 100 ml distilled water.

For the working reagent, slowly add solution 2 to solution 1. Dilute the resulting solution to 1 liter with water and mix thoroughly.

Bial's reagent (freshly prepared) for Exercise 19

Dissolve 6 g orcinol in 200 ml 95% ethyl alcohol. Then add 40 drops 10% $FeCl_3$ solution (10 g $FeCl_3$ dissolved and adjusted to 100 ml volume with distilled water). Mix thoroughly.

Biuret reagent for Exercise 27

Prepare two separate solutions:

1.   Concentrated potassium hydroxide (about 20%, w/v).
2.   0.5% (w/v) copper sulfate ($CuSO_4$).

For the working reagent, mix equal volumes of the unknown solution to be tested and concentrated KOH (solution 1). Then slowly add 1 ml 0.5% $CuSO_4$ (solution 2) and wait for color development.

Buffers for Exercise 36

Note: Other buffers are listed alphabetically according to the specific type as indicated in the Materials section of each exercise.

Sodium borate buffer (pH 8.0, 8.8, 9.0)

Prepare two separate stock solutions:

1.   0.2M boric acid: Dissolve 12.4 g boric acid and dilute to 1 liter with distilled water.
2.   0.2M sodium borate: Dissolve 76.28 g $Na_2B_4O_7 \cdot 10 H_2O$ and dilute to 1 liter with distilled water.

For buffers of different pH, combine the appropriate amounts of the above solutions as indicated below:

Buffer Number	Boric Acid (0.2M) Amount in ml	Sodium Borate (0.2M) Amount in ml	pH
1*	50	30.0	8.8
2	50	4.9	8.0
3	50	59.0	9.0

* Buffer number 1 should be used for the enzyme preparation and all reaction mixtures unless otherwise indicated.

Sodium borate buffer (pH 10.0)

Dissolve 12.4 g boric acid in 100 ml 1N (carbonate free) sodium hydroxide solution. Then dilute the resulting solution to 1 liter with distilled water.

For the buffer, add six volumes of the above solution to four volumes of 0.1N sodium hydroxide solution. The pH should be about 10.0. Use this buffer for the appropriate reaction mixture (Exercise 36, Part F).

Phosphate buffer (pH 6.2)

Prepare two separate stock solutions:

1. $Na_2HPO_4 \cdot 2H_2O$: Dissolve 11.876 g and dilute to 1 liter with distilled water.

2. KH_2PO_4: Dissolve 9.078 g and dilute to 1 liter with water.

For the buffer (pH 6.2), use two volumes of solution 1 (Na_2HPO_4) to eight volumes of solution 2. Check and adjust the pH, if necessary. Use the buffer as indicated in Exercise 36, Part F.

Citrate-phosphate buffer (pH 6.8) for Exercise 38

Prepare two separate stock solutions:

1. 0.1M citric acid (monohydrate): Dissolve 21.014 g $C_6H_8O_7 \cdot H_2O$ and dilute to 1 liter with distilled water.

2. 0.2M sodium phosphate (dibasic): Dissolve 53.614 g $NA_2HPO_4 \cdot 7H_2O$ and dilute to 1 liter with distilled water.

For the final buffer solution (pH 6.8), combine 182 ml 0.1M citric acid (solution 1) with 618 ml sodium phosphate (solution 2). Dilute the mixture to 1 liter with distilled water. Use this buffer to prepare the sodium bicarbonate solutions used in the same experiment (see Materials section, Exercise 38).

Cuprammonia reagent for Exercise 32

Prepare a saturated solution of cupric hydroxide by dissolving cupric hydroxide in concentrated ammonium hydroxide. When the solution is saturated, decant (do not filter) the dark blue solution into a glass bottle. Stopper the bottle (use a glass stopper), and store in the refrigerator. Use the reagent cold for best results.

Diphenylamine reagent (freshly prepared) for Exercise 18

Dissolve 1 g diphenylamine in 100 ml concentrated H_2SO_4. Store in a drop bottle with a glass stopper. This solution is highly corrosive, and contact with skin and clothing should be avoided.

Green safelight for Exercises 43, 45

See *Light* in this Appendix.

Iodine (I₂KI) reagent for Exercise 22

Dissolve 2.5 g I_2 and 5 g KI in 50 ml distilled water. Dissolve one component at a time and mix thoroughly. For the tests, use a 10% dilution of the above I_2KI reagent by combining 1 ml stock reagent with 9 ml distilled water.

Iodine–potassium iodide solution (I₂KI) for Exercises 2, 32

Dissolve 15 g potassium iodide in 600 ml distilled water. Dissolve into the resulting solution 3 g crystalline iodine, and adjust the final volume of solution to 1 liter with distilled water.

Light (sources and filters) for Exercises 43, 45

1.   Green safelight (transmission range of 510–550 nm): This light may be constructed from a desk lamp equipped with a 15 watt green fluorescent tube (Sylvania F15T12/G). The light should be filtered through a layer of dark green cellophane (Dennison Manufacturing Co., Maynard, Mass., generally obtainable in most art supply stores); green Plexiglas (Rohm and Haas Co., Independence Mall, West Philadelphia, Penn.); or a combination of dark green, deep amber, and medium blue cinemoid filters (Kliegl Bros., Universal Lighting Co., Inc., 32–32 48th Ave., Long Island City, N.Y. or Kliegl Bros., Western Corp., 4726 Melrose Ave., Los Angeles, Calif.).

Before use, the transmission spectrum of the filter system should be determined in a spectrophotometer (see *Light filters — transmission spectrum*). This is especially important since the characteristics of the filter media may be different from one lot to another. For example, a layer of green cellophane is not always sufficient, but the desired effect may be obtained by a combination of green and blue cellophane.

The suitable filter should then be placed in front of the bulb and taped tightly to the lamp shade so that there are no light leaks. As an alternative to using a desk lamp, it is also possible to manufacture a green safelight from a flashlight simply by covering the lens with the appropriate filter media.

2.   Far-red light (transmission range of 700–735 nm): This light may be constructed from a desk lamp equipped with a 40 or 60 watt incandescent bulb. Generally a combination of four layers of blue, one layer of green, and two layers of red cellophane (see *Green safelight* for manufacturers) will suffice as the filter system. However, the characteristics of the cellophane may be quite variable, and the transmission spectrum of the filter should be checked and adjusted, if necessary (see *Light filters — transmission spectrum*).

If a flood lamp is used, it should be separated from the sample to be irradiated by a beaker of water approximately 6 cm deep. Also, high-wattage bulbs may bleach the pigments in the cellophane with prolonged exposure. With low-wattage bulbs, the filter may be placed in front of the light source and taped tightly to the lamp shade. An alternative method is to place the sample in a light-tight box with the cover removed and replaced by the filter.

3.   Red light (transmission range of 590–700 nm): The red light source may be constructed from a desk lamp equipped with a red or white incandescent bulb (40 watt), a cool white fluorescent tube, or a red fluorescent tube. The light should be filtered through red cellophane (two layers are generally used), red Plexiglass, or a suitable cinemoid filter (see *Green safelight* for manufacturers).

To ensure the appropriate transmission range, check the filters in a spectrophotometer (see *Light filter — transmission spectrum*). Tape the filter tightly to the lamp shade in front of the bulb as stated previously (see *Far red light*). An alternative method is to place the sample to be irradiated in a suitable light-tight box with the cover removed and replaced with the filter. For convenience, a complete photobiology kit including filters and all the necessary equipment may be purchased from Carolina Biological Supply Co., Burlington, North Carolina or Powell Laboratories Division, Gladstone, Oregon. Before use, however, the filters should be checked as to their light transmission characteristics and the equipment adjusted accordingly.

4.   Light filter — transmission spectrum: To determine the transmission spectrum of the desired filter system, cut a portion of the medium to be used into a rectangle of the same height and width or diameter as a spectrophotometer cuvette.

Insert the rectangle into the cuvette on the side through which light will first pass when the cuvette is placed in the compartment of the spectrophotometer. For proper measurement of the transmission spectrum, it is absolutely essential that the filtering medium be placed flush against the side and in such a manner so that all of the light passes through the filter only once. Also, prepare a second cuvette in the same manner but with colorless or clear media.

Measure the transmission spectrum of the colored filter at 15 nm intervals from 450 to 750 nm. Be sure to adjust the spectrophotometer at 100% transmission at each wavelength with the cuvette containing the colorless media. If the selected filter system does not conform to the desired transmission spectrum, determine the approximate combinations of material by trial and error. Once the desired spectrum is obtained, the filter may be used as indicated previously.

Magnesium reagent (freshly prepared) for Exercise 18

Dissolve 0.1 g Na_3PO_4 and 30 g NH_4Cl in 50 ml distilled water. Add 2.5 ml concentrated NH_4OH and dilute to 100 ml with distilled water. Store in a drop bottle.

Methyl red solution for Exercise 2

Dissolve 0.5 g methyl red in 300 ml of 95% ethyl alcohol. Dilute this solution to 500 ml with distilled water and mix thoroughly.

Molisch reagent for Exercise 19

Dissolve 10 g α-naphthol in 100 ml 95% ethyl alcohol. Store the reagent in a glass stoppered bottle.

Nelson's arsenomolybdate reagent for Exercise 56

Prepare two separate stock solutions:

1. Dissolve 25 g ammonium molybdate $[(NH_4)_6Mo_7O_{24} \cdot 4H_2O]$ in 450 ml distilled water. Then add 21 ml concentrated sulfuric acid and mix.
2. Dissolve 3 g $Na_2AsO_4 \cdot 7H_2O$ in 25 ml distilled water.

For the working solution, combine solutions 1 and 2 and store in a brown bottle. Prepare at least 1 day before use. If the reagent develops a green tint, it is unfit for use.

Ninhydrin spray reagent (freshly prepared) for Exercise 26

Prepare a 0.2% solution of ninhydrin by dissolving 0.2 g triketohydrindene hydrate in 100 ml of water-saturated n-butanol.

p-anisidine reagent (freshly prepared) for Exercise 21

Add 2 ml phosphoric acid to 50 ml 95% ethyl alcohol. Dissolve 0.5 g p-anisidine in the mixture. Then add excess concentrated HCl until the solution turns purple.

Phenylhydrazine reagent for Exercise 19

Add 2 g phenylhydrazine hydrochloride to 30 ml water and stir. Pure compound should dissolve completely. Filter the solution, if necessary. Then add 3 g anhydrous sodium acetate and mix. (*Caution!* Phenylhydrazine is poisonous. Avoid contact with skin.)

Phosphate buffer (0.3M, pH 7.0) for Exercise 34

Dissolve 4 g KH_2PO_4 and 5.2 g K_2HPO_4 one at a time. Dilute the solution to 100 ml with distilled water. Adjust the pH to 7.0 with HCl or NaOH.

Phosphate-citrate buffer (pH 5.6) for Exercise 22

Prepare two separate stock solutions:

1. 100 ml 0.1M citric acid solution.
2. 100 ml 0.2M Na_2HPO_4 solution.

For the working buffer, mix 42 ml of citric acid (solution 1) with 58 ml sodium phosphate (solution 2). If necessary, adjust the pH to 5.6 with HCl or NaOH.

Porous clay cup for Exercise 7

This cup may be obtained from Mrs. Burton E. Livingston, Sherwood Ave., Riverwood, Baltimore, Maryland.

Potassium phosphate buffer (2mM, pH 5.8 to 6.0) for Exercise 64

Prepare two separate stock solutions:

1. 0.02M potassium phosphate (monobasic): Dissolve 0.272 g KH_2PO_4 and dilute to 100 ml with distilled water.

2. 0.02M potassium phosphate (dibasic): Dissolve 0.348 g K_2HPO_4 and dilute to 100 ml with distilled water.

For the working buffer, pour 20 ml of stock solution 1 in a beaker and adjust the pH to within the desired range (pH 5.8 to 6.0) by slowly adding small quantities of solution 2. Be sure to mix the solution and check the pH after each addition of solution 2. After adjusting the pH, dilute the resulting solution tenfold. Check the pH and use as indicated in the exercise.

Resorcinol–hydrochloric acid reagent for Exercise 40

Prepare two separate solutions:

1. Dissolve 0.1 g resorcinol and 0.25 thiourea in 100 ml glacial acetic acid. Store in a dark brown bottle.

2. Dilute five parts of concentrated HCl with one part distilled water.

For the working reagent, just before use, mix one part of the resorcinol solution (solution 1) with seven parts of the diluted HCl (solution 2).

Ruthenium red solution for Exercise 32

Prepare 100 ml of slightly alkaline distilled water by adding five drops of 10% ammonium hydroxide solution. Add just enough ruthenium red to the alkaline water to impart a light pink color to the solution. This solution may be stored in a dark bottle for a few days, but it is recommended that it be freshly prepared just before use.

Seliwanoff's reagent for Exercise 19

Dissolve 0.05 g resorcinol in 100 ml diluted HCl (concentrated $HCL:H_2O - 1:2$). Mix thoroughly.

Sodium bicarbonate (10^{-2}M and 10^{-3}M) in citrate-phosphate buffer for Exercise 38

Prepare the citrate-phosphate buffer (pH 6.8) as indicated in this Appendix (see *Citrate-phosphate buffer*).

1. 10^{-2}M sodium bicarbonate solution: Dissolve 0.420 g $NaHCO_3$ in citrate-phosphate buffer (pH 6.8). Adjust the final volume to 500 ml with buffer.

2. 10^{-3}M sodium bicarbonate solution: Dissolve 0.042 g $NaHCO_3$ in citrate-phosphate buffer (pH 6.8). Adjust the final volume to 500 ml with buffer.

Somogyi's reagent for Exercise 56

Prepare two separate solutions:

1. Dissolve 24 g anhydrous sodium carbonate (Na_2CO_3), 16 g sodium bicarbonate ($NaHCO_3$), 12 g sodium potassium tartrate, and 140 g anhydrous sodium sulfate (Na_2SO_4) in 800 ml distilled water. Be sure to dissolve completely one component before adding the next.

2. Dissolve 4 g copper sulfate ($CuSO_4 \cdot 5H_2O$) and 40 g anhydrous sodium sulfate in 200 ml distilled water.

For the working reagent, prepare just before use by combining four volumes of solution 1 with one volume of solution 2. Mix thoroughly before use.

Soybean (cultivars that may be used for tissue culture) for Exercises 59, 60

Although *Glycine max* L. 'Acme' is commonly used as the bioassay for cytokinins and is considered to be the best for sensitivity and specificity, the following cultivars can also be used for class: Agate, Flambeau, Norchiet, Grant, Blackhawk, Chippewa, Dinfield, Shelby, Clark, Peking, Dorman, Hood, Lee, Jackson, and Biloxi.

Starch solution (0.5%, w/v) for Exercise 22

Suspend 5 g soluble starch in 50 ml distilled water. After mixing, add the starch suspension to 950 ml hot distilled water. Vigorously stir the mixture and use as indicated in the exercise.

Sudan III solution for Exercise 32

Dissolve 0.1 g Sudan III into 50 ml 95% ethyl alcohol. Then stir 50 ml glycerol into the solution.

Tris (hydroxymethyl) aminomethane (THAM) buffer (0.1M pH 7.6) for Exercise 23

Dissolve the appropriate amount of THAM (about 12.1 g) in 980 ml distilled water. Adjust to pH 7.6 with HCl and adjust the final volume to 1 liter with water. Cut the proportions appropriately if less than a liter is required.

APPENDIX III

Atomic weights, common ions and formulas, and major organic functional groups and symbols

APPENDIX TABLE 2   Atomic weights

Element	Symbol	Atomic Number	Atomic Weight	Element	Symbol	Atomic Number	Atomic Weight
Actinium	Ac	89	(227)	Cobalt	Co	27	58.9332
Aluminum	Al	13	26.98154	Copper	Cu	29	63.546
Americium	Am	95	(243)	Curium	Cm	96	(247)
Antimony	Sb	51	121.75	Dysprosium	Dy	66	162.50
Argon	Ar	18	39.948	Einsteinium	Es	99	(254)
Arsenic	As	33	74.9216	Element 104		104	(261)
Astatine	At	85	(210)	Element 105		105	(262)
Barium	Ba	56	137.34	Element 106		106	(263)
Berkelium	Bk	97	(247)	Erbium	Er	68	167.26
Beryllium	Be	4	9.01218	Europium	Eu	63	151.96
Bismuth	Bi	83	208.9804	Fermium	Fm	100	(257)
Boron	B	5	10.81	Florine	F	9	18.99840
Bromine	Br	35	79.904	Francium	Fr	87	(223)
Cadmium	Cd	48	112.40	Gadolinium	Gd	64	157.25
Calcium	Ca	20	40.08	Gallium	Ga	31	69.72
Californium	Cf	98	(251)	Germanium	Ge	32	72.59
Carbon	C	6	12.011	Gold	Au	79	196.9665
Cerium	Ce	58	140.12	Hafnium	Hf	72	178.49
Cesium	Cs	55	132.9054	Helium	He	2	4.00260
Chlorine	Cl	17	35.453	Holmium	Ho	67	164.9304
Chromium	Cr	24	51.996				

Continued

APPENDIX TABLE 2   Continued

Element	Symbol	Atomic Number	Atomic Weight	Element	Symbol	Atomic Number	Atomic Weight
Hydrogen	H	1	1.0079	Protactinium	Pa	91	231.0359
Indium	In	49	114.82	Radium	Ra	88	226.0254
Iodine	I	53	126.9045	Radon	Rn	86	(222)
Iridium	Ir	77	192.22	Rhenium	Re	75	186.207
Iron	Fe	26	55.847	Rhodium	Rh	45	102.9055
Krypton	Kr	36	83.80	Rubidium	Rb	37	85.4678
Lanthanum	La	57	138.9055	Ruthenium	Ru	44	101.07
Lawrencium	Lr	103	(260)	Samarium	Sm	62	150.4
Lead	Pb	82	207.2	Scandium	Sc	21	44.9559
Lithium	Li	3	6.941	Selenium	Se	34	78.96
Lutetium	Lu	71	174.97	Silicon	Si	14	28.086
Magnesium	Mg	12	24.305	Silver	Ag	47	107.868
Manganese	Mn	25	54.9380	Sodium	Na	11	22.98977
Mendelevium	Md	101	(258)	Strontium	Sr	38	87.62
Mercury	Hg	80	200.59	Sulfur	S	16	32.06
Molybdenum	Mo	42	95.94	Tantalum	Ta	73	180.9479
Neodymium	Nd	60	144.24	Technetium	Tc	43	(97)
Neon	Ne	10	20.179	Tellurium	Te	52	127.60
Neptunium	Np	93	237.0482	Terbium	Tb	65	158.9254
Nickel	Ni	28	58.70	Thallium	Tl	81	204.37
Niobium	Nb	41	92.9064	Thorium	Th	90	232.0381
Nitrogen	N	7	14.0067	Thulium	Tm	69	168.9342
Nobelium	No	102	(225)	Tin	Sn	50	118.69
Osmium	Os	76	190.2	Titanium	Ti	22	47.90
Oxygen	O	8	15.9994	Tungsten	W	74	183.85
Palladium	Pd	46	106.4	Uranium	U	92	238.029
Phosphorus	P	15	30.97376	Vanadium	V	23	50.9414
Platinum	Pt	78	195.09	Xenon	Xe	54	131.30
Plutonium	Pu	94	(244)	Ytterbium	Yb	70	173.04
Polonium	Po	84	(209)	Yttrium	Y	39	88.9059
Potassium	K	19	39.098	Zinc	Zn	30	65.38
Praseodymium	Pr	59	140.9077	Zirconium	Zr	40	91.22
Promethium	Pm	61	(145)				

APPENDIX TABLE 3 Common positive and negative ions and their formulas

Positive Ion	Formula	Negative Ion	Formula
Aluminum	Al^{3+}	Acetate	$C_2H_3O_2^-$
Ammonium	NH_4^+	Arsenate	AsO_4^{3-}
Antimony(III)	Sb^{3+}	Bicarbonate	HCO_3^-
Antimony(V)	Sb^{5+}	(Hydrogen carbonate)	
Arsenic(III)	As^{3+}	Bisulfate	HSO_4^-
Arsenic(V)	As^{5+}	(Hydrogen sulfate)	
Barium	Ba^{2+}	Bisulfite	HSO_3^-
Bismuth(III)	Bi^{3+}	(Hydrogen sulfite)	
Cadmium	Cd^{2+}	Borate	BO_3^{3-}
Calcium	CA^{2+}	Bromate	BrO_3^-
Chromium(III)	Cr^{3+}	Bromide	Br^-
Cobalt(II)	CO^{2+}	Carbonate	CO_3^{2-}
Copper(I)	Cu^+	Chlorate	ClO_3^-
(Cuprous)		Chloride	Cl^-
Copper(II)	Cu^{2+}	Chlorite	ClO_2^-
(Cupric)		Chromate	CrO_4^{2-}
Hydrogen	H^+	Cyanide	CN^-
Iron(II)	Fe^{2+}	Dichromate	$Cr_2O_7^{2-}$
(Ferrous)		Fluoride	F^-
Iron(III)	Fe^{3+}	Hydride	H^-
(Ferric)		Hydroxide	OH^-
Lead(II)	Pb^{2+}	Hypochlorite	ClO^-
Magnesium	Mg^{2+}	Iodate	$1O_3^-$
Manganese(II)	Mn^{2+}	Iodide	I^-
Manganese(IV)	Mn^{4+}	Nitrate	NO_3^-
Mercury(II)	Hg^{2+}	Nitrite	NO_2^-
(Mercuric)		Oxalate	$C_2O_4^{2-}$
Nickel(II)	Ni^{2+}	Oxide	O^{2-}
Potassium	K^+	Perchlorate	ClO_4^-
Silver	Ag^+	Permanganate	MnO_4^-
Sodium	Na^+	Peroxide	O_2^{2-}
Tin(II)	Sn^{2+}	Phosphate	PO_4^{3-}
(Stannous)		Phosphite	PO_5^{3-}
Tin(IV)	Sn^{4+}	Silicate	SiO_2^{2-}
(Stannic)		Sulfate	SO_4^{2-}
Titanium(III)	Ti^{3+}	Sulfide	S^{2-}
(Titanous)		Sulfite	SO_3^{2-}
Titanium(IV)	Ti^{4+}	Thiocyanate	SCN^-
(Titanic)			
Zinc	Zn^{2+}		

Symbol	Name	Compound	Functional Group
—C—O—C— (with :O: on each C)	Acid anhydride	CH₃—C—O—C—CH₃ (with :O: on each C)	Acetic anhydride
—ÖH	Alcohol and phenol	CH₃CH₂—ÖH	Ethyl alcohol
—C—H (with :O:)	Aldehyde	CH₃CH₂—C—H (with :O:)	Propanal
C=C	Alkene or olefin	H₂C=CH₂ (drawn with H's)	Ethylene
—C≡C—	Alkyne	H—C≡C—H	Acetylene
—C—N— (with :O:)	Amide	CH₃—C—NH₂ (with :O:)	Acetamide
—N̈H₂	Amine, primary	CH₃—N̈H₂	Methylamine
—N̈H—	Amine, secondary	CH₃—N̈H—CH₃	Dimethylamine
—N̈—	Amine, tertiary	CH₃—N̈—CH₃ with CH₃	Trimethylamine
—C—O—H (with :O:)	Carboxylic acid	CH₃CH₂—C—Ö—H (with :O:)	Propanoic acid
—C—Ö— (with :O:)	Ester	CH₃—C—Ö—CH₃ (with :O:)	Methyl acetate
—Ö—	Ether	CH₃—Ö—CH₃	Dimethyl ether
—Ẍ:	Halide	CH₃CH₂—F̈:	Ethyl fluoride
—C— (with :O:)	Ketone	CH₃—C—CH₃ (with :O:)	Acetone
—S̈H	Thiol or mercaptan	CH₃CH₂—S̈H	Ethanethiol
—S̈—	Sulfide	CH₃CH₂—S̈—CH₂CH₃	Diethyl sulfide
—S̈—S̈—	Disulfide	CH₃—S̈—S̈—CH₃	Dimethyl disulfide
—S—ÖH (with :O: above and :O: below)	Sulfonic acid	CH₃—S—ÖH (with :O: above and :O: below)	Methanesulfonic acid

Chemical reagents, solutions of common acids, and pH values of acids and bases

APPENDIX TABLE 5   Chemical reagents

Reagent	Formula	Molecular Weight	Density 25°C (g/cc)	Solubility, 25°C (g/100 g H$_2$O)
Alcohol, 95% (ethyl)	C$_2$H$_5$OH	46.07	0.804	Miscible
Ammonium sulfate	(NH$_4$)$_2$SO$_4$	132.15	1.77 (20°)	77
Benzene, technical, for cleaning glassware	C$_6$H$_6$	78.11	0.879 (20°)	0.07 (20°)
Boric acid	H$_3$BO$_3$	61.84	1.44 (20°)	6.0
Bromocresol green (sodium salt of tetrabromo-m-cresolsulfonphthalein)	(C$_{21}$H$_{12}$O$_5$SBr$_4$)Na$_2$	742.03		Soluble
Mercuric chloride	HgCl$_2$	271.52	5.44	6.6
Phenolphthalein	C$_{20}$H$_{14}$O$_4$	318.33	1.30	0.2 (20°)
Potassium chloride	KCl	74.56	1.99	35.5
Potassium permanganate	KMnO$_4$	158.04	2.70	7.6
Potassium sulfate	K$_2$SO$_4$	174.23	2.66	10.7
Silver nitrate	AgNO$_3$	169.89	4.35 (19°)	72
Sodium acetate, granular	NaC$_2$H$_3$O$_2 \cdot$3H$_2$O	136.09	1.45	46.5 (20°)
Sodium bicarbonate	NaHCO$_3$	84.01	2.20	10.3
Sodium chloride	NaCl	58.45	2.16	35.9
Sodium hydroxide, pellets	NaOH	40.00	2.13	114

APPENDIX TABLE 6   Solutions of common acids

Reagent	Molecular Weight	Molarity	Weight Percent Reagent	Density, 20°C (g/cc)	Preparation
Acetic	60.05	17.4	99.5	1.051	Reagent-grade glacial acid
Hydrochloric	36.465	12.4	38.0	1.188	Reagent-grade concentrated HCl
Hydroflouric	20.01	25.7	45.0	1.143	Reagent-grade concentrated HF
Nitric	63.02	15.4	69.0	1.409	Reagent-grade concentrated HNO_3
Perchloric	100.47	11.6	70.0	1.668	Reagent-grade concentrated $HClO_4$
Phosphoric	98.00	14.7	85.0	1.689	Reagent-grade concentrated H_3PO_4
Sulfuric	98.08	17.6	94.0	1.831	Reagent-grade concentrated H_2SO_4

APPENDIX TABLE 7   pH values of 0.1N solutions of a variety of acids and bases*

Acids	pH Value	Bases	pH Value
Hydrochloric acid	1.0	Sodium bicarbonate	8.4
Sulfuric acid	1.2	Borax	9.2
Phosphoric acid	1.5	Ammonia	11.1
Sulfurous acid	1.5	Sodium carbonate	11.36
Acetic acid	2.9	Trisodium phosphate	12.0
Alum	3.2	Sodium metasilicate	12.2
Carbonic acid	3.8	Lime (saturated)	12.3
Boric acid	5.2	Sodium hydroxide	13.0

* Acids are listed in order of decreasing strength; bases, in order of increasing strength.

APPENDIX V

Greek alphabet, physical constants and conversion factors, numerical equivalents, and stopper sizes

APPENDIX TABLE 8   Greek alphabet

Form		Name	Form		Name
Capital	Lowercase		Capital	Lowercase	
A	α	Alpha	N	ν	Nu
B	β	Beta	Ξ	ξ	Xi
Γ	γ	Gamma	O	o	Omicron
Δ	δ	Delta	Π	π	Pi
E	ϵ	Epsilon	P	ρ	Rho
Z	ζ	Zeta	Σ	σ	Sigma
H	η	Eta	T	τ	Tau
Θ	θ	Theta	Υ	υ	Upsilon
I	ι	Iota	Φ	ϕ	Phi
K	κ	Kappa	X	χ	Chi
Λ	λ	Lambda	Ψ	ψ	Psi
M	μ	Mu	Ω	ω	Omega

APPENDIX TABLE 9   Physical constants and conversion factors

Physical Constants	Conversion Factors
Atomic mass unit	1 amu $= 1.6605 \times 10^{-24}$ g
Avogadro's number	$N = 6.0221 \times 10^{23}$ mol^{-1}
Gas constant	$R = 8.2056 \times 10^{-2}$ L atm K^{-1} mol^{-1}
Ideal gas molar volume	$V_m = 22.414$ L mol^{-1}
Masses of fundamental particles:	
Electron (e^-)	$M_e = 0.0005486$ amu $= 9.1096 \times 10^{-28}$ g
Proton (p^+)	$M_p = 1.007277$ amu $= 1.67261 \times 10^{-24}$ g
Neutron (n)	$M_n = 1.008665$ amu $= 1.67492 \times 10^{-24}$ g
Speed of light (in vacuum)	$c = 2.997925 \times 10^8$ ms^{-1}
Length	1 kilometer (km) = 0.621 miles = 10^3 meters (m)
	1 meter (m) = 39.4 inches (in.)
	= 10^2 centimeters (cm)
	= 10^3 millimeters (mm)
	= 10^6 micrometers (μm)
	= 10^9 nanometers (nm)
	1 inch (in.) = 2.54 centimeters (cm)
	1 Angstrom (Å) = 10^{-8} centimeters (cm)
Volume	1 cubic meter (m^3) = 10^3 liters (l) = 264.2 gallons (gal)
	1 liter (l) = 10^3 milliliter (ml)
	= 10^3 cubic centimeters (cm^3)
	= 1.06 quarts (qt)
	1 milliliter (ml) = 1 cubic centimeter (cm^3)
	1 gallon (gal) = 4 quarts (qt)
	1 quart (qt) = 2 pints (pt) = 946 milliliters (ml)
Mass	1 kilogram (kg) = 10^3 grams (g) = 2.20 pounds (lb)
	1 pound (lb) = 16 ounces (oz) = 454 grams (g)
	1 ounce (oz) = 28.4 gram (g)
Energy	1 kilocalorie (kcal) = 10^3 calories (cal) = 4.18 kilojoules (kJ)
	1 calorie (cal) = 4.18 joules (J)
	1 British thermal unit (BTU) = 1055 joules (J)
Pressure	1 atmosphere (atm) = 760 millimeters of mercury (mm Hg)
	= 760 Torr = 101.3 kilopascals (kPa)
Temperature	$°C = \dfrac{(°F - 32) \times 5}{9}$
	$°F = \dfrac{°C \times 9}{5} + 32$
	$K = °C + 273$

APPENDIX TABLE 10   Numerical equivalents

Unit	Equivalent
Angstrom	0.1 nm
Inch	25.4 mm
Foot	0.3048 m
Yard	0.9144 m
Mile	1.60934 km
Square inch	645.16 mm^2
Square foot	0.092903 m^2
Square yard	0.836127 m^2
Square mile	2.58999 km^2
Cubic inch	16.3871 cm^3
Cubic foot	28.3168 dm^3
U.S. fluid quart	946.3 cm^3
Avoirdupois (pound)	0.45359237 kg

APPENDIX TABLE 11   Stopper sizes

Stopper Size Number	Tubes (mm OD)	Openings (mm ID)	Top Diameter	Bottom Diameter	Length
00	12–15	10–13	15	10	26
0	16–18	13–15	17	13	26
1	19–20	15–17	19	15	26
2	20–21	16–18.5	20	16	26
3	22–24	18–21	24	28	26
4	25–26	20–23	26	20	26
5	27–28	23–25	27	23	26
5½	28–29	25–26	29	25	26
6	29–30	26–27	32	26	26
6½	30–34	27–31.5	34	27	26
7	35–38	30–34	37	30	26
8	38–41	33–37	41	33	26
9	41–45	37–41	45	37	26
10	45–50	42–46	50	42	26
10½	48–51	45–47	53	45	26
11	52–56	48–51.5	56	48	26
11½	57–61	51–56	60	51	26
12	62–64	54–59	64	54	26
13	64–68	58–63	67	58	26
13½	68–75	61–70	75	61	35
14	80–90	75–85	90	75	39
15	92–100	83–95	103	83	39

APPENDIX VI

Supplementary readings

Becker, W. M. 1977. *Energy and the Living Cell: An Introduction to Bioenergetics.* New York: Harper and Row, Inc.

Bewley, J. D. and M. Black. 1978. *Physiology and Biochemistry of Seeds. I. Development, Germination and Growth.* Berlin: Springer-Verlag.

Clayton, R. K. 1980. *Photosynthesis: Physical Mechanism and Chemical Patterns.* Cambridge: Cambridge University Press.

Crozier, A. E. (ed.). 1983. *The Biochemistry and Physiology of Gibberellins.* New York: Praeger Scientific.

Devlin, R. M. and F. H. Witham. 1983. *Plant Physiology*, 4th edition. Boston: Willard Grant Press.

Epstein, E. 1972. *Mineral Nutrition of Plants: Principles and Perspectives.* New York: Wiley.

Evans, D. A., W. R. Sharp, P. V. Ammirato, and Y. Yamada. 1983. *Handbook of Plant Tissue Culture, Vol. 1.* New York: Macmillan Publishing Co., Inc.

Evans, D. A., W. R. Sharp, P. V. Ammirato, and Y. Yamada. 1984. *Handbook of Plant Tissue Culture, Vols. 2 and 3.* New York: Macmillan Publishing Co., Inc.

Galston, A. W., P. J. Davies, and R. Satter. 1980. *The Life of the Green Plant*, 3rd edition. Englewood Cliffs, N.J.: Prentice-Hall, Inc.

Giese, A. C. 1979. *Cell Physiology.* Philadelphia: W. B. Saunders Co.

Goodwin, T. W. and E. I. Mercer. 1982. *Introduction to Plant Biochemistry*, 2nd edition. New York: Pergamon Press.

Haliwell, B. 1981. *Chloroplast Metabolism.* Oxford: Oxford University Press.

Hall, J. L. and D. A. Baker. 1978. *Cell Membranes and Ion Transport.* New York: Longman, Limited.

Jacobs, W. P. 1979. *Plant Hormones and Plant Development.* Cambridge: Cambridge University Press.

Khan, A. A. (ed.). 1982. *The Physiology and Biochemistry of Seed Development, Dormancy and Germination.* Amsterdam: Elsevier Biomedical Press.

Kramer, P. J. 1983. *Water Relations of Plants.* New York: Academic Press.

Lehninger, A. L. 1982. *Principles of Biochemistry.* New York: Worth.

Letham, D. S., P. B. Goodwin, and T. J. V. Higgins (eds.). 1978. *Phytohormones and Related Compounds: A Comprehensive Treatise.* Amsterdam: Elsevier/North-Holland Biomedical Press.

Lüttge, U. and N. Higinbotham. 1979. *Transport in Plants.* New York: Springer-Verlag.

Mayer, A. M. and A. Poljakoff-Mayber. 1980. *The Germination of Seeds,* 3rd edition. Oxford: Pergamon Press.

Milburn, J. A. 1979. *Water Flow in Plants.* New York: Longman, Limited.

Mohr, H. 1978. *Lectures on Photomorphogenesis.* New York: Springer-Verlag.

Moorby, J. 1981. *Transport System in Plants.* New York: Longman Group, Limited.

Moore, T. C. 1979. *Biochemistry and Physiology of Plant Hormones.* New York: Springer-Verlag.

Nicholls, D. G. 1981. *Bioenergetics: An Introduction to the Chemiosmotic Theory.* New York: Academic Press.

Nobel, P. 1983. *Biophysical Plant Physiology and Ecology.* San Francisco: W. H. Freeman and Company.

Noggle, G. R. and G. J. Fritz. 1983. *Introductory Plant Physiology,* 2nd edition. Englewood Cliffs, N.J.: Prentice-Hall, Inc.

Raven, P. H., R. F. Evert, and H. Curtis. 1981. *Biology of Plants,* 3rd edition. New York: Worth Publishers, Inc.

Salisbury, F. B. and C. W. Ross. 1985. *Plant Physiology,* 3rd edition. Belmont, Cal.: Wadsworth Publishing Company, Inc.

Siegelmann, H. W. and G. Hind (eds.). 1978. *Photosynthetic Carbon Assimilation.* New York: Plenum Press.

Stryer, L. 1981. *Biochemistry,* 2nd edition. San Francisco: W. H. Freeman and Company.

Ting, I. P. 1982. *Plant Physiology.* Reading, Mass.: Addison-Wesley Publishing Company, Inc.

Wareing, P. F. and I. D. J. Phillips. 1978. *The Control of Growth and Differentiation in Plants,* 2nd edition. New York: Pergamon Press.

Wilkins, M. B. (ed.). 1984. *Advanced Plant Physiology.* London: Pitman Publishing, Limited.

Index